Screenwriting With a Conscience

Ethics for Screenwriters

LEA's COMMUNICATION SERIES
Jennings Bryant / Dolf Zillmann, General Editors

Selected titles include:

Berger • Planning Strategic Interaction: Attaining Goals Through Communicative Action

Ellis • Crafting Society: Ethnicity, Class, and Communication Theory

Greene • Message Production: Advances in Communication Theory

Heath/Bryant • Human Communication Theory and Research: Concepts, Contexts, and Challenges, Second Edition

Olson • Hollywood Planet: Global Media and the Competitive Advantage of Narrative Transparency

Penman • Constructing Communicating: Looking to a Future

Perry • American Pragmatism and Communication Research

Salwen/Stacks • An Integrated Approach to Communication Theory and Research

For a complete list of titles in LEA's Communication Series, please contact Lawrence Erlbaum Associates, Publishers

Screenwriting
With a Conscience
Ethics for Screenwriters

Marilyn Beker
Loyola Marymount University

2004

LAWRENCE ERLBAUM ASSOCIATES, PUBLISHERS
Mahwah, New Jersey   London

#51537168

Lawrence Erlbaum Associates, Inc., Publishers
10 Industrial Avenue
Mahwah, NJ 07430

Cover design by Kathryn Houghtaling Lacey

Cover concept by Marilyn Beker

Library of Congress Cataloging-in-Publication Data

Beker, Marilyn.
   Screenwriting with a conscience : ethics for screenwriters /
   by Marilyn Beker.
       p.   cm.
Includes bibliographical references and index.
ISBN 0-8058-4127-X (alk. paper)
ISBN 0-8058-4128-8 (pbk. : alk. paper)
   1. Motion picture authorship—Moral and ethical aspects. I. Title.

PN1996.B453 2004
174'.982—dc21
                                                    2002041677
                                                        CIP

FOR

My Parents
Joe and Bronia Beker
who taught me the value of integrity and truth

AND

My Husband
Greg Rorabaugh
whose profound goodness, unwavering sincerity
and unconditional love
inspire me to live with
gratitude, faith, and courage.

*Change yourself and you have done
your part in changing the world. Every
individual must change his own life
if he wants to live in a peaceful world.
The world cannot become peaceful unless
and until you yourself begin to work
toward peace.*

—*Paramahansa Yogananda*

CONTENTS

PART V: KILLING THE MESSENGER

PART VI: HAVING WRITTEN AND WRITING MORE

CONCLUSION

PREFACE

WHY ETHICS? WHY ME? WHY NOW?

The most remarkable person I ever knew had the power to change lives and profoundly influence how people thought. He wasn't famous and he didn't work in media. In fact, he lived quietly in the middle of a forest. When I met him, he was 78 years old but he had the appearance, vitality, and attitude of a much younger man. In fact, he was so energetic he would often outwork his 22-year-old assistants!

J. Oliver Black had been wildly successful in life. He was a husband, father, major industrialist, multimillionaire, and the largest private land owner in Michigan. But after his wife died and his children were grown, he gave up everything to do with business and glamor to live a life of spiritual introspection and philanthropy. And yet, in spite of his renunciation of materialism, he was a profoundly practical person and exceptionally joyful.

His joy came from self-realization and a devotion and commitment to high ideals, to his core beliefs and values. That joy drew people to him, encouraged them, and inspired them to transform themselves. "It's a great life if you don't weaken," he'd say, and urged people not to give in to the bad habits, wrong thinking, weak will power, and negative environment that might weaken them. He was a great believer in living well and to him that meant living with integrity, purpose, and personal responsibility.

That certainly is a great way to live but it's not an easy thing to do—particularly in this age of profound paradox. Our world is on the cutting edge of technology; we've made great breakthroughs in medicine and art; we've worked the social sciences to profound degrees and excelled in creating the finest communication systems ever. And yet, in spite of all the great and good ideas brought to function, we haven't been able to rid the planet of poverty, disease, and war.

We haven't been able to eliminate even minor misunderstandings between people—misunderstandings that often severely disrupt personal peace. We may not even be able to understand ourselves enough to eliminate the personal stumbling blocks keeping us from reaching our own potential.

How many times have we been out of touch with our own feelings, our own beliefs, our own intentions? And even when we were in touch with our feelings, beliefs, and intentions, how many times have we tried to make them known to someone close to us—even someone we loved deeply and who deeply loved us—and been misunderstood? How many times have our sincere attempts at honesty been mistaken for guile or our humor for sarcasm? How many times have we tried to play it cool only to be taken for an uncaring clod or, worse, an ineffectual nerd?

No matter how hard we try, it becomes quickly clear that communicating with another human being (even one from our own culture, background, and belief system) is one of the most difficult things we can do. It gets hard when we're not sure exactly what we want to say and even harder when we succumb to the confusion that results from the paradoxical demands society makes upon us.

Our North American society encourages us to exercise our democratic right to express ourselves but at the same time, for the sake of decorum, demands that our expressions don't offend. We're expected to say things that make us seem sensitive and caring when we interact with others, while we are encouraged to do things that could be construed as economically and socially savage when we do business or involve ourselves in our careers. We're pushed to compete and win in every arena going; we're measured by stringent external standards; we judge each other harshly; and yet, in public at least, we must appear to conform to the prevailing sentiments of the day if we want to get ahead.

School teachers who don't believe in bilingual education are afraid to say so for fear of being called racist; parents who want to make sure their kids remain bilingual are afraid to insist on bilingual education for fear they will be called un-American; shopkeepers anxious to attract business don't want homeless people sleeping in their doorways but are told that the "rights" of the homeless must be protected even if the exercise of those "rights" keeps business away; people who are homeless feel disenfranchised and unheard; employers are afraid to fire inefficient employees for fear of being sued; employees are afraid to blow the whistle on unscrupulous employers for fear of losing their jobs; schoolchildren who want to exercise their rights of free speech by talking about God in the classroom are forbidden to do so because that might offend people who don't believe in God; people who don't believe in God are offended by the Pledge of Allegiance.

Wherever we turn for guidance on what to do or say, we get mixed and often incomprehensible messages. We aren't sure just how much we can say before we get sued, fired, shunned by neighbors, dropped by friends, or attacked by strangers. Ours has become a culture of tightrope walkers negotiating a precarious razor's edge of political correctness. That's why it takes so much courage to speak our mind and to stand up for our beliefs.

The prevailing sentiment seems to condone subterfuge over candor. No wonder executives at WorldCom and Enron Corporations (belly-up in 2002) thought nothing about hiding the truth of their companies' impending financial disasters. Although those corporate debacles made big news, people didn't seem particularly surprised that savvy corporate CEOs siphoned off big personal bucks before their companies crashed and financially ruined trusting shareholders and employees.

And if the public tends to suspect that greed and corruption are a part of corporate big business, it absolutely takes for granted that greed and corruption are an integral part of show business. That's because almost everyone realizes that in show business, the competition even to get work, let alone make it to the top, is ferocious, and brutal.

In Hollywood, a town high on hype, people who work in show business are expected to be tough enough to take punishing rejection, brazen enough to get noticed, ruthless enough to win out over the strongest competition in the world, and at the same time make us all believe they're likeable, modest, considerate, and just plain folks.

Desperate to work, to be commercially viable, to be a household word, many in the popular creative arts have chosen to defend their right (and everyone else's) to do and say anything to get ahead. The code of Hollywood is survival. In the entertainment industry, like in the oceans, the Amazon rain forest, and the world's wildlife preserves, it's all about eating or being eaten.

Fueled by this sense of desperation and fear ("I won't work if I say the wrong or unpopular thing"), most screenwriters have bought into an "anything goes" attitude. Falling hard for the notion that to be successful they have to give everyone's point of view respect, they don't respect themselves enough to stand up for what they believe. And that's because many of them have for so long altered their screenplays to satisfy other people and have sold out so often by giving in to the pressures of the market place that they no longer know what their beliefs are.

Well, enough is enough! I'm tired of people laughing when I say it's possible for screenwriters to have ethics. I'm tired of hearing that ethics for screenwriters is an oxymoron. I know there are lots of screenwriters who are ethical and believe that ethics should be an essential ingredient of how we treat each other, how we work, and what we choose to write. It's time for those of us who believe in ethics to go public.

Personally, ethics have been important to me since childhood. My parents impressed on me that honesty, integrity, and fairness were not only laudable characteristics of a "good" person but were also essential to peace of mind. In fact, my parents (probably because of their life-shattering experiences in Eastern Europe during World War II) firmly believed and imparted to me that ethics were cherished ideals—a matter of honor and soul survival in the face of any adversity. That's probably why, when I made up my mind to be a writer at age 10, I determined with the idealism of youth that everything I wrote would be ethical at its core.

Since then, I've learned that things aren't that simple. In all the forms of professional writing I've done (poetry, short stories, a novel, magazine pieces, news, radio and television documentaries, children's programs, TV miniseries and screenplays), I've had the opportunity to observe and experience in myself, as well as in others, the difficult and sometimes complicated process of making ethical choices in the writing life. From these observations and experiences I've been able to develop my own personal code of ethics—ethical standards I fight to uphold; standards that help me draw lines in the sand when it comes to making artistic and practical decisions. Ethics still matter a great deal to me and I always make a conscious and concerted effort to demonstrate that in my life and in my writing.

I especially try to do that when I teach. Since 1984, I've been a screenwriting professor at Loyola Marymount University in Los Angeles. Our students have won lots of Emmys, a Sundance Award, and even an Oscar for screenwriting. Our School of Film and Television is a competitive and first-class arena where I've worked hard to make sure my students know the nuts and bolts of the screenwriting craft and even harder to inculcate in them a respect for the power their images will wield on screen.

At the same time, in deference to my commitment to the ethics of freedom of expression and the First Amendment, I've been very careful to respect my students' "space" and their points of view. I've worked with them to make their scripts better, to make them work, no matter what the message of those scripts, because I believed that it wasn't my job to interfere with the students' right of expression even if the ideas expressed rankled or offended me.

Over the years, though, I've become more and more troubled by the kinds of screenplays my students are writing. "Inspired" by what they see in the theaters, many of their preoccupations have been without purpose or meaning and many demonstrated a marked lack of social concern. I've grown tired of students telling me they've included horrific violence and over-the-top sex because they want studio executives to notice them; of students trying to convince me that graphic scenes of women being beaten and tortured are just what their target audience of 18- to 25-year-old males would pay to see on a Friday night.

Finally, in the face of an escalation of social debacles—Columbine and its clones, increased crime, drug use and sexual activity among children, irresponsible television pandering to the most base human motives ("reality" television shows like *Who Wants to Marry a Millionaire, Survivor, Temptation Island*), and a seeming growing disinterest in humanistic standards on the part of entertainment executives—I determined that, as a writer and an educator preparing screenwriters for the professional arena, I could no longer grit my teeth and say that every screenwriter's point of view and value system, no matter what, has to be championed in the name of free expression.

I'm sick of reading hopefully commercial scripts with little or no social conscience, inhumane points of view, degrading philosophies, senseless acts of graphic violence. As audiences, artists, or educators I think it's time we all drew an ethical line in the sand that says (like the Peter Finch character in *Network*) "We're mad as hell and we're not going to take it anymore!"

I know that because of the current preoccupation with civil liberties, it's wildly unpopular in the artistic community to exercise any kind of control on creative work. I know that many of my screenwriting colleagues might be shocked that I believe certain forms of expression shouldn't be welcomed into the popular entertainment arena. But in the interests of humanity and peace, I've decided that there must be a turn toward some kinds of standards in the popular creative arts if entertainment is to uplift and inspire audiences to take on the problems of life.

I believe firmly that inspiration and upliftment that provide a new way of looking at life is the true purpose of popular art and entertainment. Of course, it is possible that inspiration and upliftment can come by examining the "dark side" of life, but only if that examination is intended to provoke thought and not solely to shock and titillate.

Even this view may be unpopular. To many, entertainment is simply diversion—a distraction away from the "reality" of every day. But more and more the stuff intended for diversion is becoming our "everyday reality."

Many Americans live lives that are fraught with violence or the threat of violence. The media plays up the impression that our city streets are unsafe and that danger lurks everywhere. It's hard to achieve peace of mind and a feeling of safety and security with images of violence bombarding us. I believe that every human being ultimately wants to live in a peaceful society filled with thoughtful people. That's why it's incumbent on each of us to do what we can to contribute to the peace and harmony of the world. Screenwriters, because they are the first line in the creation of the images in popular entertainment—images that profoundly affect society—have the opportunity and profound responsibility to make a huge contribution. The processes described in this book are designed to help screenwriters work toward this goal. In creating work based firmly on an ethical platform, screenwriters can make an important step toward improving the human condition while still entertaining audiences and realizing their own ambitions and dreams.

—*Marilyn Beker*

INTRODUCTION

THINKING BEFORE WRITING

This book is for the thinking screenwriter—the person who is serious about his or her craft and wants to maintain the integrity of his or her intentions. Many people who want to write for film and television spend most of their time thinking of stories, characters, or situations they can turn into scripts. They'll spend lots of time thinking about structure, about dialogue, about characters, and about how to sell their work. But very few will spend any time at all thinking about why they want to write, what they have to say when they do, and how they'll hold on to everything they hold dear in the screenwriting process.

That kind of thinking is the most important thinking of all for a writer to do. It's the kind of thinking (and planning) that goes before serious activities like jumping off a cliff, riding rapids, climbing a mountain, or taking a job. It's the kind of thinking that helps you decide if you want to spend an important and significant amount of time in your life doing something strenuous, difficult, and risky. Because writing a screenplay on spec (without assurance of selling it) is just that.

A good screenplay is very difficult to write, takes up a good chunk of your time and energy, only has an audience when it's made, and sometimes won't make you a red cent. And once your screenplay is made, you'll be accountable for its message, meaning, and impact on audiences. Even more daunting, the film on which your screenplay is based will become part of a body of work that will follow you around forever.

That's why it's so important for you as a screenwriter to be proud of your work. And if you know why you are writing, what you want to say, and how to say it, you'll be able to keep smiling and take any heat that might come with failure or success.

Nearly all screenwriting books concentrate on the technical aspects of screenwriting: structure, plot development, character development. This book is different because it concentrates on work the writer must do before actually writing—preparatory work that is absolutely essential in order to create a good screenplay.

Shakespeare (in *Hamlet*) wrote, "to thine own self be true and it must follow, as the night the day, thou canst not then be false to any man." For our purposes, I take this to mean that a writer must be true to herself with a capital S. The Self I am talking about is the one that knows innately, intuitively, when what is being said comes from truth with a capital T and from integrity; from positive humanistic concerns; from a place determined to benefit humanity and reaffirm life. If a writer writes from that level of Truth to Self and belief in that Truth, then that writing will ring true to an audience. Even audience members who disagree with the ideas in that kind of Truth will respect the writer for having the integrity and courage to express it.

LOOKING AHEAD: THE PROCESS

I've broken the thinking process I've talked about into six parts.

Part I: Why?

We'll begin with a practical definition of ethics and some basic guidelines for this book. Keep in mind that although this is a book on ethics for screenwriters, it includes writing techniques that will stress ethical points of view in screenplays, personal anecdotes, and industry examples and tips. That's because it's important to develop the practicality of an ethical approach as it applies to screenwriting.

That application relates directly to the reality of message and meaning in screenplays and to the writer's personal motives for screenwriting. We'll examine the reality and importance of messages and those things that motivate screenwriters. Then we'll explore the artist's social responsibility and the relationship of that responsibility to the notion of art itself.

Part II: The Certainty of What

We'll examine the screenwriter's commitment to social responsibility and to personal ethical beliefs by relating that commitment to the willingness (or reluctance) to write scenes of graphic sex and/or violence.

This examination will be directly related to the issue of conscience and its place in screenwriting. Once conscience is examined then it can be used to determine what ethical position in society the screenwriter might choose to take.

Part III: What Really Matters

As we continue to examine society, the screenwriter is invited to enter into a process of defining good and evil and given tools with which to make ethical choices. These choices are demonstrated through analysis of models from life and movies.

Part IV: White Hats, Black Hats

Here's where we look at some practical writing techniques that can be used to create "good" and "bad" characters who embody our definitions of good and evil (arrived at in Part III) and make definite choices (based on the models we've looked at). We learn how to write "ethical biographies" for our characters and how to portray dastardly deeds and angelic acts with conviction and verve.

And now that we're clear on ethical specifics and ethical standards, we'll examine how these standards might morph in genres like comedy and satire.

Part V: Killing the Messenger

Our newfound zeal for ethics will be tempered by looking at ways to maintain the integrity of these ethics without sermonizing or preaching.

Part VI: Having Written and Writing More

Now that we've worked on our writing, it's time to deal with the ethical aspects of the business—agreements, contracts, writing partnerships, and industry relationships.

In this section, we give industry anecdotes and pointers for ethical survival as a screenwriter in Hollywood.

By the time you're finished doing the exercises in the book and taking careful stock of yourself, you should have a clear understanding of what makes you tick as a writer, what is important to you, what you want to say, and how you want to say it. And, you'll have developed your own personal code of ethics as applied to screenwriting and even to life.

A PRACTICAL DEFINITION OF ETHICS

What do I mean by ethics? Books have been written hammering out definitions and even philosophies of ethics but I like to keep our "definition" simple. I take ethics to mean a code of behavior (as it applies to screenwriters, a way of presenting material and ideas) that is fair, just, honest, and integrity-based; that intentionally hurts no one; that works toward the greatest common good without destroying individuals in the process; that is honorable, consistent, and thoughtful, and that is life-affirming and humanistic.

Are ethics universal or do they change from culture to culture? There are some ethical notions and behaviors that vary from culture to culture. For example, in some cultures (notably Afghanistan and Saudi Arabia) it's considered ethical for male family members to make all decisions for women in that family—especially in legal matters. Although culturally this behavior might seem "ethical" to someone in Afghanistan, such behavior doesn't correspond with our Western notion that each individual has the right to make her own decisions and should be legally responsible for them.

Here's another example from a story that appeared in *The International New York Times* on July 23, 2002. Fadime Sahindal was a Kurd who immigrated to Sweden with her family. Rather than enter into an arranged marriage like her sisters, Fadime chose to date a Swedish man. Her father and brother hated her because of this and threatened to kill her. Fadime went to the police, who prosecuted her brother and father and fined them. A few days later, her brother (who had a criminal record for theft and drugs) attacked Fadime in the street and beat her up badly. In court he testified that Fadime was "a whore" and when confronted by his own legal history and asked if that history didn't dishonor his family, he said "I've broken your rules but Fadime has broken our rules and our rules are much more important."

Fadime's father expressed his shame over his daughter's behavior and even though he pledged not to harm her, a few months later he shot his daughter dead in front of her mother and two sisters. They did nothing to stop it and after her death one of the older sisters phoned a male member of the family and said "The whore is dead now."

In the culture specific to Fadime's family, an "honor" killing might be highly ethical. But in Swedish culture (and in North American, Western European, and other cultures) usurping of women's rights is unethical and "honor" killing is an abomination.

Personally, I believe that some ethical principals should be universal. Cultural proclivities might explain ethical variations but most often they don't excuse or nullify obvious ethical inequities (especially having to do with murder, genocide, bigotry, and subjugation).

In many ways, and at the core of what it means to be human, there are certain ethical principles that ensure the survival of the race and provide a feeling of well-being and prosperity to its members. Anthropology has shown that certain social systems are based on the ethical principles of sharing, non violence, and a reverence for life. It would seem that these principles, whether we like them or not, are at some level embedded in all of us so deeply that they are part of our species profile, the emotional genome that defines our humanity.

There are certain ethical behaviors that seem to travel well from culture to culture: the protection and love of children, respect and love for parents, reluctance to betray or cheat family members and friends, the cherishing and valuing of human life. Some might argue with Darwin that survival still comes down to competition and the triumph of the fittest, but whereas that may be true on a primitive level, on a cultural and idealistic level, perhaps the kind of survival that is most worthwhile is based on the ethical principles that provide human beings with meaning and hope.

It might not be worth surviving if we have to destroy others to survive. It might not be worth surviving if we debase ourselves and others in order to get what we want. And it might not be worth surviving if, in times of crisis, we revert to behavior that might be more "animalistic" than ideal.

Survival should mean more than just "staying alive"—making it in a ruthless system. Survival should mean the ability to thrive and expand; to feel joy, to live without guilt, and to have the self-satisfaction that comes from knowing that we did the "right" thing when there was something to be done—that we held fast to what we believed was a "righteous" ethical code.

Most of us like to think we live with integrity. From childhood, we've been taught the Golden Rule: do unto others as you would have them do unto you. By the time we're grown up we know that just as we don't want to be unfairly accused, we need to shy away from accusing others unfairly; just as we don't want to be betrayed, we need to refrain from betraying.

Of course, there are those people who don't care what's done to them, just so long as they can do something to others. People like that tend not to think of the future, don't believe that everything that goes around comes around, and do believe that they will somehow be able to protect themselves from the effects of their own bad actions.

In fact, that's seldom true. Have you ever done something you knew was wrong and had it backfire? Have you ever gossiped about anyone only to have that gossip come back to hurt you? Sometimes the hurt we experience after we do something unethical might only be internal–manifesting as feelings of guilt. But those feelings might be bad enough to make us feel horrible about our lives and about ourselves—horrible enough to keep us from enjoying the fruits of our unethical actions.

As I've already implied, I believe that each one of us knows innately when we're being honest and what is just and fair, but sometimes, because of the stakes/rewards presented and the temptations that exist all around us, we give in and behave dishonestly, unjustly, unfairly—in short, unethically. It's this giving-in that we've got to fight, and the only way in which we can fight it is to be aware of our own personal code of ethics and make a strong personal commitment to it. That way, because of our awareness of and commitment to ethics, we'll have the strength to resist temptations to do and say unethical things.

THIS BOOK'S ETHICS

Before we begin the process of becoming more aware of our ethics and learning how to apply them to screenwriting, here are the "givens" of this book.

∽ Censorship is unacceptable. There is no excuse for government, religious, or societal control imposed on creative work. Rather, screenwriters should exercise self-restraint in order to respect audiences and themselves. In fact, if, as the Dalai Lama says in *Ethics for the New Millennium*, "self-restraint is the basis of all ethical behavior," it should also be the basis for screenwriting. Self-restraint is expected in matters of craft (relating to form, execution, and structure). It should be expected in content as well.

∽ Screenwriters, because they work in a medium that has the potential to influence and affect large numbers of people, should strive to make people think about the culture in which they live, and to make statements that are based in a profound respect for humanity and the human condition.

∽ Screenwriters who want to live in peaceful societies have a responsibility to make sure their work contributes to the peace of those societies. Psychological degradation and physical violence, by themselves, do not solve problems. As Nobel Laureate Aleksandr I. Solzhenitsyn said "The fight for peace is only part of the writer's duties to society. Not one little bit less important is the fight for social justice and for the strengthening of spiritual

values in his contemporaries. This, and nowhere else, is where the effective defense of peace must begin—with the defense of spiritual values in the soul of every human being."[1]
I take spiritual to mean the way the Dalai Lama uses it, as some level of concern for others' well-being, "concerned with those qualities of the human spirit—such as love and compassion, patience, tolerance, forgiveness, contentment, a sense of responsibility, a sense of harmony—which bring happiness to both self and others."[2]

❧ Ethical screenwriting means putting forth ideas and actions that show respect for audiences, humanity, and Self. Although ethical films may contain material that is controversial or edgy, such films use this material to make a point and not for its own sake simply to excite or encourage others to perform debasing or chaotic actions.

❧ Morality is different from ethics. Morality changes with the times. What may be acceptable in one age may be unacceptable in another. In the 1800s, women's ankles were considered erotic and women who bared them were fast and loose. Today nearly bare bosoms are flaunted on fashion runways, films, television shows, magazines, and even city streets. In fact, morality fluctuates considerably in conjunction with what's going on in the world. It also changes from culture to culture. In some Muslim countries, for example, it's considered immoral for women to show their faces. In polygamous cultures, it's not considered immoral to take more than one wife.

❧ Morality is often based on church doctrine. Ethics, although arguably connected to religion, can be considered essentially humanistic. For example, a prostitute may be immoral but she can be ethical. That is, she may act in ways contrary to "church decency" by selling her body, but she is an ethical person if she honestly provides her service without ripping off people. This is an important distinction. Because it is impossible to make sure no one is offended by something and because morality is often personal and sometimes even cultural, screenwriters

---

[1]Solzhenitsyn, A. *Letter to Komoto Sedze*, Moscow, 15 Nov., 1956. Pub. *Oak and the Calf*, Harper and Row, New York, 1980, p. 458.
[2]Lama, Dalai. *Ethics For the New Millennium*, Riverhead Books, New York, 1999, pp. 22–23.

who want to include controversial material must first examine
their motives for doing so.

      Everything that appears in a screenplay should be considered in
context and by the seeming intent of the screenwriter. If material
seems to be there strictly to shock or offend, it is not ethical. If it
is clearly there to support the telling of the story and the thesis to
which the screenwriter subscribes, it will be ethical even if it de-
picts immorality. Screenwriters who make irresponsible and
thoughtless statements that will denigrate, inflame, or offend
audiences needlessly, and even inspire them to dastardly deeds,
are writing immorally and unethically.

      This means that screenwriters need to examine the things they
value—their "values" and the values of their cultures. Sometimes the values of the screenwriter might not be the values of
his own culture. In that case, he can be revolutionary by con-
demning the values of his own society. A screenwriter may or
may not, for example, value ethics and morality. People have all
kinds of different values. The word "values" alone does not necessarily mean goodness or idealism but rather what individuals hold dear and care deeply about.

      Those trying to learn the craft of screenwriting must first exam-
ine their own values and philosophies and own up to them before they create works that will shape and/or influence the
values, philosophies, and the lives of others.

    This book will help you get to the core of your story by first finding
your own core. This book will help you discover your artistic vision by
finding out what you really want to say about life and saying it in a way
that is "humanistically" responsible.

    If you believe you should be able to write whatever you want to with-
out thinking about it, if you believe anything goes in movies and that,
for the sake of big bucks, you have the right to inflict any and all of your
ideas on an unsuspecting public, this book is not for you.

    But if you truly want to contribute to society, if humanity and peace
and making a difference are important to you, if you long to make
statements about the human condition in order to uplift and inspire

others to search for truth and meaning in their own lives, then welcome! You're in for a great adventure!

WARNING: THIS BOOK CONTAINS SPOILERS!

I hate to ruin movies for people who haven't seen them by giving away the plot and ending but I've had to do that with some of the movies I've used as examples in this book. Here's a list of films I've "spoiled." You might want to see them before you read further.

Amadeus
A Simple Plan
Bandits
Boys From Brazil
Casablanca
The Contender
Double Indemnity
Erin Brokovich
Fargo
The Fugitive
Heist
It Happened One Night
Jackie Brown
Jurassic Park I and III
Knock on Any Door
Left Behind
Mission Impossible II
The Omega Code
On the Waterfront
The Prophecy
Rat Race
The Score
Strangers on a Train
Thelma and Louise
Thirteen Days
Titanic

MOVIES VERSUS TELEVISION

This book can and does apply to all writers because it is designed to help them find out what they really want to say and how to make ethical decisions in the writing process. Primarily though, it concentrates on screenwriting for movies. That includes movies made specifically for television.

It doesn't go into other kinds of television writing like situation comedies, news, documentaries, reality shows, game shows, and variety specials. That's because these forms of television are team written.

People who write in television writing teams usually have only one ethical decision to make—they need to decide whether or not they want to accept a writing position on a certain show. Before a writer is hired onto a television show, that show has already been carefully defined and all of its creative parameters delineated. If a writer doesn't believe in the show's concept or feels that it is unethical, that writer shouldn't work for that show. It's a simple decision.

Once a writer accepts a position on a television show, that acceptance implies that he or she agrees to abide by the ethics and parameters of that show. Any ensuing writing decisions that might occur in the show's process will be made collectively, first by the writing team, and then ultimately by the producers and the networks.

In movie writing, the producers and studios also have the power to change or direct the ethical integrity of the project but the individual writer has much more say, particularly in the initial stages of the writing. That's why, although I will occasionally mention television projects (particularly docudramas and movies of the week), we'll confine most of our discussion to writing movies where, at least in first draft, screenwriters get to make all the decisions, from story idea to final execution.

Whereas it's true that studio movies are audience tested and geared to marketing demands, it's still possible for a writer to be entrepreneurial and get a movie made based on a speculative original script. Given current television realities, that isn't really possible. Where it is (rare television movies made directly for cable), this book will definitely apply.

PART I

WHY??

ETHICS? FOR SCREENWRITERS????

In an industry more cutthroat than a tracheotomy, can a standard of ethical behavior be defined and met? Of course not! Screenwriters are not (as your parents persist in warning you) doctors or lawyers. *However*, there is a way of working that will make you feel good about yourself and let you sleep, day or night; a way of conducting yourself and your work by standards that command respect because they are founded on motives that are humanistic and integrity-based.

Being an ethical screenwriter isn't easy, but consider the alternative. Every day kind and wonderful people who want to behave ethically find themselves turning into ruthless wackos playing craps with their souls in the neon lottery that is the movie business. The thing that drives these people, whether they like to admit it or not, is the idea that when they get their movie made, it will mesmerize millions and make them famous and filthy rich.

You're probably saying you're not like that—the only thing you're interested in is getting your personal artistic vision across. You may demonstrate that you despise even the idea of big bucks and good tables at restaurants by dressing like a derelict and swearing allegiance to Chef Boyardee. Fine. But just how much you want to see your movie get made may warp your priorities enough so that in the end you'll do just about anything to get what you want—eat anchovies, wear a name tag, blow off your "artistic vision" to become just another shallow money grubber hurling your soul at the devil.

And just what is "artistic vision" anyway? Most screenwriters can't put theirs into words. They usually just mumble something about being able to say exactly what they want to when they want to. The sad

15

truth is that most screenwriters have no idea why they are writing or what they want to say when they do.

I've been writing professionally for over 30 years and I've met only a small a capella choir of screenwriters who have thought deeply about life and about what comments their movies will make about it. The rest are just shrugging their shoulders and hoping that "entertainment" doesn't have to be too deep.

The thing is that what we call "entertainment" is one of the deepest things around because, usually, even without us realizing it, entertainment stirs our emotions, piques our intellectual interest, and otherwise provides the stimulus that makes us think. "Entertainment" can actually transform us and change our lives.

Just think for a moment about the films that made a difference in your life. I'll bet that every one of them struck a chord in you that moved you in some way to react in the world, to come to certain conclusions about life, to at least think about how you were living.

Exercise

Take a few moments to list the films that made a difference in your life. Beside each film title write down (in one or two lines) the story of the film and then, in another line, write down what you think the film is *really* about—what message it delivers for you. Go deep—beyond the superficial messages—(e.g., good-looking guys get more sex) to deeper issues (e.g., Society rewards appearance while it overlooks soul-worth). Here are a couple of my old favorites.

The Wizard of Oz

ABOUT: A girl from Kansas is caught up in a tornado and carried to a magical land where she must outwit a wicked witch and find the wizard who can help her get back home again.

MESSAGE:There's no place like home—no matter how fabulous and exciting other places may be.

Casablanca

ABOUT: A rakish ex-pat saloon owner in WWII Casablanca meets up with the woman who broke his heart in Paris and discovers she is the wife of a revered Nazi fighter whose life he has the chance to save.

MESSAGE:War and love inspire people to act with selfless courage.

You get the idea.

An Extra Exercise

Here's a list of twelve Oscar-winning screenplays written directly for the screen. Try the About/Really About exercise on these. If you want more winners find them on the Oscar web site: www.Oscars.org. If you haven't seen the films, see them!

2001 Gosford Park—Julian Fellowes
2000 Almost Famous—Cameron Crowe
1999 American Beauty—Alan Ball
1998 Shakespeare in Love—Marc Norman, Tom Stoppard
1997 Good Will Hunting—Ben Affleck, Matt Damon
1996 Fargo—Ethan Coen, Joel Coen
1995 The Usual Suspects—Christopher McQuarrie
1994 Pulp Fiction—Roger Avary, Quentin Tarantino
1993 The Piano—Jane Campion
1992 The Crying Game—Neil Jordan
1991 Thelma and Louise—Callie Khouri
1990 Ghost—Bruce Joel Rubin

MESSAGE AND MEANING

ou'll notice that usually, what the film means to you is quite different from "what it is about." Writers, producers, directors, and screenwriting teachers will ask you what your film is about, and, usually, they just want you to tell them your story. But as writers, we all must ultimately face the truth that the story is only a chiffon scarf over the rippling bosom of its message.

Of course, old films are easier to decipher. Often they articulate their messages boldly in the body of the film. In the *Wizard of Oz* Dorothy declares "There's no place like home!" at the end of the film, and in *Casablanca*, Rick delivers his famous "hill of beans" speech on the foggy, soggy runway. But these days our film messages are more and more subtle because modern audiences don't like to be told what to think in obvious and often cheesy ways. We've got to sneak our messages into our movies if we want to deliver them, and sometimes we do such a good job of sneaking that we hide what the film is Really About even from ourselves.

You may be surprised at how difficult it was to articulate why certain films meant something to you. That's because it takes some deep soul searching to discover what's really important to you in life. Maybe now's the time you should consider what you want to say in the film you currently want to write. You may even begin to ask yourself why you want to write films at all. We'll get to that later.

But for now, the initial step is to realize that all movies say something even if the person who wrote them (and/or made them) is conscious of that or not. And all movies make a strong statement, even if that statement is silly, inconsequential, or just plain dumb (party till you throw up for no good reason; boff your brains out for no good reason; shoot, kill, maim for no good reason.) I'm sure you can find lots of movies to fit those examples.

So isn't superficiality enough? Didn't Marshal McLuhan, years ago, say that the way you say something is really what you are saying—that you can't separate the message from the medium (the medium IS the message)? Well, yes, but it's not that simple. I first met McLuhan at the University of Toronto where he was a stodgy professor and I was a poetry-crazed freshwoman. He used to appear around campus wearing scruffy tweed jackets and sensible sturdy shoes. He didn't look as hip as his ideas—an unlikely medium for his own particular message.

In the summer of my sophomore year, while working as a cub reporter for *The Toronto Star*, I interviewed him about his ideas. McLuhan reluctantly agreed to meet me in a phone booth at an old Woolworth's store. And although the location was intriguing, his conversation wasn't.

I couldn't understand a word he said. He spoke in monosyllables and cryptic weird phrases that really meant nothing and I finally dissolved in a puddle of giggles, assuming that he was giving me an example of "blip culture"—something we now call "sound bites"—a taste of some kind of meaningless future shorthand speak. I took away from that meeting the idea that maybe McLuhan was telling me the future of communication lay in the replication of electronic blips whose sole function was to fascinate by sound nuance; that if we continued at our present rate, conversation (live or electronic) would become only aural stimulation whose purpose was to titillate rather than inform.

And by rendering the example of that kind of communication, McLuhan himself was showing up the absurdity of pop culture's direction—style over substance—and making the point that communication can mesmerize and titillate but that it should also have meaning beyond glitzy technological displays.

I left our meeting delighted by his approach, captivated by his "performance," and dedicated to the ideal that meaning was important in conversation, in media, and in art. Even though I'm a big fan of bizarre image placement, ultramodern techno-installations, and image manipulation (wonderfully done in *Natural Born Killers*, a real example of style over substance), I still believe that communication should have meaning and I am convinced that McLuhan did, too.

Right now you might be getting a little nervous—afraid that I'm going to spend many chapters trying to prove to you that art and particularly movies make a difference to society, have a power over it, and sometimes inform and shape it. Relax. I'm not. There are other books out there that do that. I'm also not going to try to convince you that images are important and compelling. There are *lots* of other books that do that. I'm going to assume that you've read all those books or have chosen to ignore them because, like me, you already believe that film is magical, powerful, and can change and/or influence lives; that film has meaning and can inspire people and motivate them to action.

It's because I believe that images are powerful, meaningful and mighty things, that I also believe that the people who create them on such a large scale (a 35-foot idea is hard to ignore) have a responsibility as part of the human family to wield their power wisely. I probably got that belief early on when I began as a cub reporter.

In those days, we weren't allowed to show pictures of dead people in the newspaper. It was obvious to us that pictures in print were compelling and could in fact be insightful. Television was even more sacrosanct. I still remember watching in horror the now-famous televised execution of the Vietnamese man in the madras shirt—a direct shot to the head—in black and white on the 6 o'clock news. The broadcast of that image changed journalism forever and it changed popular culture because from that point on, it somehow became acceptable to show graphic violence on the news to audiences because they were somehow removed from it by virtue of the technology.

And yet, the people working in news—the reporters, the cameramen (then it was mostly camera men, particularly in Vietnam) working with the newest and most portable equipment in history (video cameras and videotape instead of cumbersome film equipment and film)—could somehow be involved in the event but not of it. That very Yogic concept of being in the world but not of it seemed to indicate that it was now all right to enter into more intriguing realms of image projection heretofore considered too rough for ordinary folk.

The cameraperson's responsibility in news is to record the events, not influence them or stop them. Based on that, plenty of artists have held up humanity's horror to the public gaze indicating that they were simply "reporting" on life and had no responsibility to influence or alter it. And yet journalistic responsibility is very important to audiences. Journalists are held to high standards of truth and accountability. Reporters are supposed to be objective, although reportage can never be truly objective—the reporter's very presence in the event alters that event.

The reporter maintains that he or she deals in truth. Artists maintain that they deal in truth too—a nonobjective and personal truth—but the artist certainly believes that the artistic impression is just as important as the reporter's impression, sometimes even more important. In that way, artists—writers, painters, actors—particularly those who work in the popular culture—are the examiners, interpreters, shapers, or guardians of that culture just as much as journalists are.

Screenwriters certainly belong in the category of popular artists who examine, interpret, and even shape culture, so screenwriters should own up to that responsibility in some of the ways society expects journalists to do, by admitting their work has meaning and power; by accepting that this work has an effect on society; by admitting that the messages at the core of that work come directly from the person who generated it.

Screenwriters need to be message-conscious as well as eloquent, just as reporters should be. Screenwriters need to be honest with themselves and with audiences and, by virtue of their ability to make thoughts visual and "real," screenwriters should take responsibility for what they say and dedicate themselves to putting forth material that will not harm humanity.

WHAT ABOUT THE FIRST AMENDMENT?

Should every screenwriter put forth only pleasing lollipop images? Of course not. Often it is necessary to portray ugliness to express beauty. Shouldn't writers have the right to say anything they want to even if it hurts humanity? Unfortunately, because I am a staunch supporter of the First Amendment, I believe that they should have that right. But I also believe that people should think before speaking and certainly before writing to make sure what they are saying is really what they believe. Often that isn't the case. Usually people don't really know what they believe and don't realize what the movies they are writing are really saying.

Also, people interested in protecting the First Amendment are always talking about choice. I agree it's important that people be given the choice to see something or not to see it. But, that choice must be informed. I also believe that artists, when they work, should make informed choices, too—about the possible effects of their work and about their audiences—and often they don't.

Screenwriters may say they want freedom of expression but do they really exercise it when they get the chance? Just what do we all mean by artistic freedom? Certainly, as writers, we need to think that we are free to write whatever we want to write—especially if we live in free countries where expression is valued, cherished, protected. And yet, in the final analysis, how much of what we say is influenced by what our friends will say, what our parents will say, what our peers will say about our work? How many of us are influenced by our own insecurities and embarrassments? How many of us are influenced by economy, by the demands of the market place?

There are lots of examples to draw from. Writers have told me that they won't write graphic sex scenes because they are embarrassed that their parents will see their stuff. I've seen writers make their scripts politically correct for fear of condemnation from the public. And I've known many writers who have changed the most vital points of their screen plays, moved off their own "sacred" convictions, to please studio executives and producers.

Maybe our attachment to artistic freedom is only something we tell ourselves so we can feel better. Maybe, if, as radical political philosopher Herbert Marcuse said "art is the escape valve of society," it is also the escape valve of artists who want to believe that they live in a world apart from other beings, apart from lesser mortals who must pay particular social dues.

The truth is that artists always exercise control over their work and often the inclination to go against the social grain becomes just as narrow a constraint as adhering to it.

Perhaps it's time for artists (and particularly screenwriters) to acknowledge self-control and exercise it.

If they do, they may deflect the elements in our culture—religious, social, or governmental (stronger now then they have been in the last few decades)—from forcing the implementation of controls. And controls (a loose term for censorship) and government "influence" in art can be odious, counterproductive, and even dangerous.

As a case study, let's take a consulting project I took on for the Northern Service of the Canadian Broadcasting Corporation. In 1977, I was sent to the Canadian Arctic to report on the effects of television in Arctic communities and to make suggestions for new approaches to media programming. Years before the Canadian government had set up television transmitters north of the 60th parallel and was piping in a potpourri of television shows to the native population. There were Canadian documentaries and educational programs but there were also American prime time reruns of *Hawaii Five-O* and *The Streets of San Francisco*.

Each night, small TVs were hooked up to generators in northern government prefab houses, and communities of Inuit would watch guys in suits shoot each other. At first the Inuit were confused by TV. They couldn't figure out editing and wanted to know how a guy could change clothes so fast and get from one place to another without walking.

Northern native audiences assumed that the South was filled with powerful magicians who could appear and disappear at will. At first, they even mourned TV deaths and were shocked at the violence. But as time wore on, interesting things began to happen to Inuit communities. The guns that the Inuits had used only for hunting began to be used to inflict violence. In fact, violent crimes—usually scarce in Arctic communities—increased sharply. Children learned English from TV and stopped speaking their native languages. They began to disrespect the tribal elders. They lost interest in practicing the old arts, crafts, and survival skills.

As it imported television, the government set up stores to supply Arctic communities with easily available munchables. Gradually, hunters stopped hunting and the whole population relied more and more heavily on government assistance. As the government piped in more and more culture by televison, junk food, cigarettes, and booze, the population's spiritual, mental, and physical health deteriorated. I saw kids as young as 7 smoking cigarettes. And the native spokespersons who recognized the problem told me what caused it.

When the government sent stuff up to the Arctic, the people assumed it was good. Their original native culture had been based on the fact that the tribal elders knew what was best for the community and would not allow anything bad into it. The people now looked on the government as an elder and assumed that it was doing what was best for the people.

Cigarettes couldn't be bad, nor could sugar or TV, in which guns were used to kill people. The community was being destroyed by the new influx of what was supposedly "good for them" according to government decree, just as much as it was being destroyed by the violent images on American TV.

It's clear from this example (I'm sure you've got some of your own), that government doesn't necessarily know what's good for us. So who does? Maybe we do! Most people would agree that everyone longs for happiness, for a sense of comfort and well-being. Screenwriters, like other artists and most people, want to live in communities that are safe and pleasant. Most artists are not interested in the destruction of society. In exercising self-control, artists not only protect society, they protect themselves.

After working on probably a thousand scripts, I've come to realize that most screenwriters really want to enrich humanity, and I fervently hope that the bulk of people in the world are basically good (how un-Hollywood of me to say so) and would act accordingly if made conscious of the effect of their actions. I really do think that those who write what some would call socially inappropriate, violent, or debased scripts would back off if they realized what their movies were actually saying and had to publicly take responsibility for those messages.

So how can you as a screenwriter who believes that images are powerful, magical, and effective, and who wants to live safely and happily in a pleasant world and to contribute to humanity rather than denigrate it, get your messages across, maintain your integrity, and keep inspiration strong and vital? The following chapters will give you some practical ways to do that.

It'll take work on your part but if you are willing to get to know your characters, to explore what motivates them, to give them depth and nuance, you should be more than willing to do the same for yourself. You come before the characters, before the story, before everything, and knowing what you are all about is the real measure of success in show business as in any business, as in life.

THE CERTAINTY OF WHY

"GOTTA DANCE!!!"

—*Arthur Freed and Nacio Brown via Gene Kelly,*
Singing in the Rain (1952)

n *Singing in the Rain* (1953 WGA Award for Best Written American Musical), geeky dance wannabe Gene Kelly arrives in New York looking for stardom. And no matter what happens, no matter what trials confront him, he keeps on singing just one thing: "Gotta dance!!!" That's always seemed to me the perfect reason for taking the kind of brutality that show business has to offer. You simply just "gotta" dance, write, direct, act, whatever. The fact that you just "gotta" is the thing that drives you and will get you through any rough spot. "Gotta" is like an incurable disease you can't shake, like an addiction that no amount of "cold turkey" can cure. It's a love that never goes bad.

But, let's face it, most people who write screenplays don't really "gotta." From close observation and personal experience I have come to realize that most people write screenplays because they want fame, money, material rewards, travel, respect from their peers, awe from the thugs who bullied them in college, and sex from practically anyone. It should be clear to everyone that "gotta score" isn't quite the same as "gotta write movies." In fact, it's drastically different.

The kind of "gotta!" in the "gotta write" I'm talking about is based on the firm belief that the writer has something to say and must, at any cost, say it. Writers who "gotta write" will keep on writing even though no one buys their stuff, even though everyone calls them crazy, even though friends leave them, and they wind up mumbling to themselves in the dark. Writers who "gotta," simply love the act of writing and they must write or die.

The quintessential example of that kind of writer is Alexander Solzhenitsyn, Russian writer, intellectual, and Nobel prize winner. In his literary memoir *The Oak and The Calf* (1980), Solzhenitsyn chronicled, with heartbreaking precision, the agony he endured to make sure his writing survived the nightmarish censorial regime of Cold War Russia.

Imprisoned in the early 1950s for seditious thinking, Solzhenitsyn spent 2 years in the Soviet Gulag—prisons and concentration camps where a single line could cost him his life. During that time, he wrote nothing down and memorized his work as he composed it. Here's how he did it:

> *I improvised decimal counting beads and, in transit prisons, broke up matchsticks and used the fragments as tallies. As I approached the end of my sentence I grew more confident of my powers of memory and began writing down and memorizing prose—dialogue at first, but then, bit by bit, whole densely written passages. My memory found room for them! It worked. But more and more and more of my time—in the end as much as one week every month—went into the regular repetition of all I had memorized."[3]*

Once released from the Gulag, but still in exile, he was told that he had cancer and had only 3 weeks to live. Terrified that all he had memorized in the camps would be lost, he hurriedly copied everything he remembered in tiny handwriting, rolled the thin sheets of paper into cylinders and hid them in a champagne bottle that he buried in his garden.

But he did not die from his cancer and that "miracle" as he called it, gave him the fresh energy to continue writing in secret during the remainder of his exile even though he was forbidden by the government to do so. As he wrote, he instantly destroyed the rough copies and drafts. He preserved the final drafts by creating small hiding places for the manuscripts, some of which he microfilmed and then smuggled out of the country.

Even after his exile, when he moved to Central Russia, he continued to live in fear that he would be found out and punished for his subversive creativity. He continued to devise hiding places.

> *I wrote on especially thin paper, destroyed outright all rough drafts, outlines and superseded versions. I typed as tightly as possible, leaving no space between lines and using both sides of the paper; and by burning the fair copy of the manuscript as soon as the copying was*

---

[3]Solzhenitsyn, A. *The Oak and the Calf*, Harper and Row, New York, 1980, p.3

*finished. This method was followed for my novel "The First Circle,"
my story "ShCH-854," and my film script Tanks Know the Truth.*[4]

Solzhenitsyn lived this way for 13 years and finally, he lived to see
his work published and to win that Nobel Prize. Of course, his story is
quite remarkable and extreme. But it demonstrates admirably how
much someone wholly dedicated to truth will champion that truth in
spite of any obstacle. "For the writer intent on truth," he wrote, "life
never was, never is (and never will be!) easy: his like have suffered every imaginable harassment—defamation, duels, a shattered family
life, financial ruin or lifelong unrelieved poverty, the madhouse, jail.
While those who wanted for nothing, have suffered worse torments in
the claws of conscience."[5]

Solzhenitsyn surely is the poster boy for "gotta" when it comes to
writing, and few of us could put ourselves in that same category.
Happily, those of us who live in America don't have to. Theoretically,
we can write what we want to write and not worry about being jailed.
And yet, even in America, there's a subtle tyranny at work that can be
every bit as oppressive as the Russian Gulag if we let it!

What I'm talking about is the intimidation we face by writing even
though that writing may never see the light of day. Most of us are afraid of
the " terrible suffering" that Solzhenitsyn outlines in the foregoing paragraph—suffering having to do with the absence of money, fame, and respect. Back to why most screenwriters are writing in the first place.

Solzhenitsyn urged us to remember that there is a suffering more
terrible than poverty and obscurity. That suffering is caused by lying
to oneself, by harboring a guilty conscience, by knowing that one is a
coward or suspecting that one might be. It is the suffering endured by
those who are afraid that they are shams with nothing of real importance to say. It is suffering caused by the confusion of inconclusive decisions and vague personal principles and beliefs. Solzhenitsyn
pointed out that those who write because they truly feel a message inside them burning to be heard, tend to feel their suffering less. At least
they can sleep at night knowing that they are being true to themselves
and their own visions.

A writer with a message will fight harder against censorship and repression than one who is uncommitted to a point of view. And that
fight will be much more effective because it is arguable and defendable, based not only against the principle of censorship, but *for* a personal and definite belief.

---

[4]Ibid., p. 5
[5]Ibid., p. 3.

To most Americans, the mere idea of censorship is odious . We trea-
sure a free society and, in fact, these days perhaps take it a little for
granted. Interestingly, Solzhenitsyn's book *The Oak and the Calf*, so
popular when it first came out in the 1970s, is no longer in print today
and can be found only through used book services. Perhaps its rail
against censorship seems outdated now. And yet I believe it should be
reissued if only as a reminder of the dangers of repression. Few students are old enough to remember the ravages of the Soviet regime
and its effects.

I suppose I feel so strongly about it because I saw it at work in my
own life. I had an uncle who survived the Holocaust in Europe by flee-
ing to Russia. He was stuck behind the Iron Curtain after the war, forbidden to leave or travel. By torturous means, my parents tracked him
down and began a correspondence with him. For over 20 years, until
the Soviets finally allowed him to leave, my uncle wrote letters to us.
They arrived infrequently and we opened them with trembling hands
only to find that his words to us had been obliterated by huge blocks of
dark black ink. The government censors had done their work.

Because I was a child, I imagined that my uncle must have tried to
write us some deep dark secrets about the government. But after my
uncle arrived in Canada in the early 1970s, he told us that he'd only
written mundane things in the letters—the size of his house, what he
was doing for fun, what work he did, and so on. Nothing that we
thought was earthshattering or important. Apparently the Soviet government did not want even that innocuous information imparted to
the free world. And my uncle was writing only to one small family.
Imagine then how strictly censorious such a regime would be to those
writing for the general public.

I believe none of us want our government censoring our private letters. And certainly it follows that we wouldn't want it to intervene in
our more public venues, interfering with our right to know. Of course,
letter writing is different from more public forms of writing. If you
were writing a personal letter, you might say things that you wouldn't
say to large groups of people.

In fact, most people who write letters intended for public consump-
tion—for example, letters to the editor—usually take a great deal of
time making sure that those letters clearly reflect their public
thoughts. And yet, sad to say that screenwriters who are writing to a
great many people indeed and getting their message across to millions, seldom admit that they are putting forward any messages at all.

I think that's because so few screenwriters are held accountable for
their work. The reality of the industry is such that all writers are
re-written, often by throngs of other writers, to fit the studio standards. Most writers say they often don't even recognize their work

when it gets to the screen. In many cases, writers denigrate their own work (particularly true in television) in order to distance themselves from studio and network changes.

At Writer's Guild events, I've often heard writers call their last project "a piece of crap" and roll their eyes and then laugh when they talk about what they are working on. It's almost as if they're saying, "Hey, I know the show's garbage but I'm taking the money and running."

It's always interesting to me when these same writers vehemently condemn censorship. They claim they want to be able to say whatever they want to in their scripts but sadly, most of them never even get close to knowing what that is. The truth may be that they're less afraid of censorship than they are of introspection.

That's because introspection is hard—even harder than writing—but I believe it must be the first step in any project. First draft is where writers get to say exactly what they want to say and the amount they're willing to change their scripts clearly indicates where their allegiances lie and what truly motivates them. I'll get into that later, but first let's look at the process of introspection.

Introspection is not an easy thing to do. It is in fact the basis of many religions that urge people to get calm and examine their own thoughts. It is the cornerstone of meditation. And also, although some are unwilling to admit it, it really is also the foundation of the artistic process.

The idea that they have to think before they write, that they actually have to say something, makes some writers really anxious. It's not always fun to sit down, go deep inside of yourself, and face your truth head on. But it is always fascinating, and can be exciting and even therapeutic.

It's unfortunate that our schools don't teach the introspection process as thoroughly as they do other processes. We are taught how to put things together, how to compute, how to construct. In the arts, we teach people process as far as structure and materials are concerned. But seldom do we teach students how to make artistic decisions that are based on inner values instead of physical constructs, material limitations, or form restrictions.

Many teachers of artistic subjects (often artists themselves) are afraid to deal with content because they don't want to get too personal, but I believe there is nothing more personal than teaching an artist how to do art. Art education should be based on personal introspection and it needs to begin in grade school. Unfortunately, it seldom does. Most often it's left to college or graduate school and even there it's given short shrift.

Look at the curricula of most film schools. They all dwell on process but precious few of them ever include classes on artistic and social issues, the psychology of art, or personal value systems. There's a

lot to be said about how to do a thing but few people who educate filmmakers and screenwriters bother to delve into the scary world of what
to write about. Educators just assume that people who want to be in
the arts know what they want to say about life and about themselves.

I suppose that's because people believe that a filmmaker's point of
view is his or her own business. Well, professionally it's not. It's something that affects lots of people. And unfortunately most of us are not
born with the capability of expressing our point of view artistically.
That has to be learned. And, because a point of view is based on personal values, these need to be explored and discovered.

In addition, if writing is a process of making artistic decisions
based on a personal point of view stemming from personal values (see
Wonder Boys for the film explanation of that process), how do we educate ourselves about making artistic decisions? If parents don't do it
because they believe schools will and schools don't have time in their
busy curricula to do it, then where do we learn how to make artistic
decisions based on personal values?

It seems that the only way to do that is to first educate ourselves
about our own personal value system by engaging in concentration
and introspection. Once we do that, we can come to our own conclusions and make sure the decisions we make are not based on what
someone else tells us to do, upon commercial appeal, the mood of the
day, or laziness.

In the following chapters we explore ways of determining our own
value system. I know that it's not an easy process and it takes years—a
whole life time really. The great thing is that we don't have to wait to
write until we have figured it all out. Writing is in fact one of the ways in
which we do figure it out. That's why, sometimes, the process is so difficult if it's done well. It's easier to write the common pap of the day
than to squeeze what you really feel about something onto the empty
screen of your computer.

Because that process is tough and sometimes painful, many
screenwriters say that they hate the act of writing. This is a mystery
to me. Why would you want to do what it takes if you hate, even despise, doing it? If writing causes you pain, stop writing! It's simple
enough. I'd say the same thing to someone banging his head against a
chrome dinette set.

The sad truth is that the rewards for writing screenplays—
self-discovery and artistic satisfaction—are found often in the very act
of writing itself. That's where the real fun is. That's where you learn
about yourself, your family, your community, your world. That's
where you discover the thrill of creating something outside of yourself
that speaks for you.

So now back to your motivation for writing! Do you have something you want to find out and something you want to say or would you rather spend time selling, changing, refuting, fighting, disclaiming, and forgetting your work?

Are you in it for the money? It can be paltry when you spread it out over the years it takes to get a project off the ground. Those screenwriters lucky enough to get their work optioned are chagrined to learn the fees are meager ($1,500 for a year; $500 for six months) or even nonexistent. Lots of writers have optioned scripts for a buck to big companies they hope will have the clout to get their projects made.

And even then a writer might take years to sell anything and when she finally does, that first project will probably sell for Guild minimum (2000 figures: Low $45,400: High $85,330). The writer might then end up in rewrites for years.

A student of mine sold a script to a burgeoning production company for about $60,000. It sounded like big money at the time and he blew lots of it on toys and parties. But 4 years later he was still doing rewrites for the company and hoping to get the movie made. Divide 4 into 60 and voilá ... he was making about $15,000 a year. Not as good a Mcjob.

And then there are the screenwriters who work Guild-free and down and dirty trying to get a movie made even if it's someone else's vision. I've got several students writing scripts for rich guys who can't write but want to get into the biz or for straight-to-video movie mills. They make $5,000 a pop. This can seem like a lot to a starving student, but that figure loses its luster after graduation when churning out maybe four or five scripts a year grosses you $25,000. There are easier ways to make a better living!

You're probably not eager to believe those realities. You may firmly believe you'll be the one to beat the statistics, to break the bank, to hit the long shot. Okay. That's not a bad thing. There is a certain strength and courage to be found in naivete that can keep you going when the nights are particularly black. So let's get down to finding out how important a motivator money really is to you.

To do that, give it a percent figure. Is it 100% of your reason for writing screenplays? 50%, 30%? Now realize that percentage is how much you will be willing to compromise your message to get your script sold. If your figure was around 30%, you'll probably spend most of your "career" feeling misunderstood, unappreciated, and angry at the world for not paying you to let you write exactly what you want to without interference. If your figure was around 50%, you're going to spend a lot of time being torn between your need to express yourself and your desire to make money. If your figure was around 75% or more, know that you'll find yourself compromising so much that you won't recognize yourself or your script at the end of the day. If that's the case, you might be better off going into manufacturing. If you think you can get rich giving people what they want, then you should choose a business where people actually know what they want (selling shoes or waffles) because none of the people in Hollywood really know what they want. They say they do but they do not. And so jumping higher and higher to please those kinds of people may get you a spot on the Olympic team but it won't necessarily get you sales or money.

Exercise

Make a list of what you want in life. Include as much as possible. Don't make things broad. You can include nonmaterial things. This list will change, of course. Decide which of those things you can get by doing something else. Do that something else and leave screenwriting alone.

But maybe you don't care much about money. Are you in screenwriting mostly for the fame? Ha! Nobody ever thinks of the writer. There may be two or three out there that people know but it's a director's medium and no one ever lets you forget that by never remembering you—especially at parties. Ask your nonmovie-addicted friends to name five screen writers. Their vacant stares should tell you something about the fame you can expect. And if people ask you what you do and you tell them you're a screenwriter, watch out for the inevitable question: "Have you written anything I might have seen?" If you say no, their initial enthusiasm turns to disdain usually reserved for the terminally pitiful.

Do you want the respect of your peers and awe from bullies? Can friends really respect you when you keep canceling dates on them in favor of preparing for pitch meetings or trying to shmooze phonies who say they can get you in the door? Do you want people to base their respect for you primarily for being a name in a long list of credits?

And bullies? There are more of them in show business than there ever were in your neighborhood playground. They beat you to a creative pulp and then try to muffle your screams by offering you money and dummy credits. More often though, they get you to do your best work, rip it from your sweaty naive hands, and stomp on it with cleated jack boots. Here's an example from my personal Hollywood Horror Vault of what you can expect.

I wrote a miniseries for a fledgling producer with big connections. We partnered up on the project—I did all the writing and grunt work in exchange for her door opening. My partner got us both in to see the VP in charge of a major studio who was powerful enough to get the project made at one of the networks. We walked into the meeting, sat down, made small talk, the VP complimented me on my writing, and then said to my partner, "Why do you need Marilyn? Bump her off the project. You'll get more money and we can do what we want to it. Get rid of her." I sat there with my mouth open. Had they no shame? No consideration for my feelings? No respect for me?? Of course not.

I tried to fight back by looking brave and telling them I could take care of myself, but the damage had been done. By the end of the meeting, the partner who had promised to stick by me "no matter what" had jammed herself into the last lifeboat and was paddling toward the rescue ship and away from me at breakneck speed.

Paul Attanasio, Oscar and WGA writing award nominee for *Quiz Show* and *Donnie Brasco*, and creator and writer of *Homicide: Life on the Street* and *Gideon's Crossing*, explains Hollywood horror by saying, "It's just all about the money; it's all about the business, exclusively. There's so little consciousness that we're promulgating a uniquely American art form. Now the distribution wings are running whole studios. They have their formulas and figures, and that's what it's about. It's just nearly impossible to do anything good."[6]

Of course, these kinds of statements and stories usually have little or no effect on the young writer anxious to take that bull of show business by the horns. The cruel realities of the biz are ignored by those who believe their passion makes them immune to Hollywood hurts! What do I mean by passion? Wanting something so much that you hurt; that you're willing to take risks, make sacrifices, look foolish, fail trying. If you're passionate about something it's always at the back of your mind, you're always thinking about it, you're always making plans. The truly passionate never take no for an answer and believe in the inevitability of realizing their dream.

Well, if you're one of these people and you actually do have passion then consider this: Too much passion is also a dangerous thing. How dangerous? Put a percentage on the amount passion motivates you. If your passion level is over 75%, know that you probably won't be willing to compromise enough to get your movie made! Passion for your message can be a strong motivator but it can also be a real deal breaker.

This is where it gets ugly, even for altruistic you. The trick is this: You have to be willing to give up enough of your principles to sell your movie but not enough to make you feel like a cheap whore. That doesn't mean

---

[6]Waldman, Alan, At Play in the Fields of Film and TV (a conversation with Paul Atttanasio), in *Written By: The Magazine of the Writer's Guild of America*, West, April 2001, p.23.

rationalizing to yourself that big bucks will make you feel better about selling out. It means that you have to know just how far you are willing to compromise your message by changing it, making it more subtle, or altering it.

As an example, let's take the abortion issue. Are you willing to concede certain points on abortion? Let's say you're against it. How much against it? Is it okay in cases of rape or incest? Is it okay if it threatens the mother's life? If it's never okay for you and the producer wants you to say it sometimes is, would you be willing to move off your position?

Remember that when they asked William Faulkner (who was a writer on *To Have and Have Not*, *The Big Sleep*, and *Gunga Din*) why he wrote for movies instead of continuing to write books (*The Sound and The Fury*, *Light in August*, *As I Lay Dying*), he said he'd rather get part of his message across to 12 million people than all of it across to 12 thousand. Know what you want to get across and what you're willing to give up to do it. (We'll tackle how to find what you want to get across in a later chapter.)

Every successful writer in Hollywood will tell you that each moment is a compromise, that filmmaking is a collaboration, that your work will be altered—sometimes beyond recognition—at every stage. The only draft that is really your own is the first draft, so you've got to take your stand there and declare your passion. You can move off it later if you're willing to.

If you want a little peek at that process, read William Goldman's *Adventures in the Screen Trade*, and take a look at *The Big Picture* with Kevin Bacon, *Swimming With Sharks* with Kevin Spacey, and *The Player* with Tim Robbins. Those insider views of Hollywood demonstrate how no one is safe from that process. Even Oscar Winner Brian Helgeland said in *The Los Angeles Times* "First Person" feature (May 14, 2001) "Movie making is a fight. Or a beating. And when the director, the producer and the studio all have a different idea of what the film should be, there's going to be some sort of fight." So if you're going to fight, make sure you know what you're fighting for and that you're passionate about it.

And now for the dark side of motivation. There's a little thing about being too motivated. And that little thing, called persistence by some, can blow itself way out of proportion. Motivation can be so entrenched that for years struggling screenwriters unwilling to "sell out" by compromising their messages, or unwilling to give up the idea of themselves as a screenwriter, can end up losing marriages, relationships, and other well-paying and promising careers because they are so driven by the need to sell a script.

Here's where "gotta dance" can end up zinging you. Life usually requires you to be practical. It's one thing to sit around a garret eating soda crackers and ketchup soup when you're in your 20s and single. It's quite another to require your kids and spouse to do that.

If you're lucky enough to get a significant other who supports you, you can drag it on for awhile as she works while you write, but know that resentment will eventually float into the relationship and sink you. Unfortunately many nonwriters have the sense that "writing" is not really working. They see selling as working. In fact, the Writers Guild even looks at it that way. The number of writers who "work" each year is recorded in sales figures. No one counts the hours and hours of backbreaking computer time you're putting in your little cubby behind the water heater. Sad and tragic, but true.

So, give yourself some reasonable time in which to succeed before you pack it in and get another job that will pay the rent and put braces on the kids' teeth. I'm not talking here about giving up or giving in. I'm talking about surviving. I find 5 years a nice rule of thumb. If you haven't sold anything in 5 years, the chances are that things are grinding too slowly for you. Now that doesn't mean that you don't have talent, what it takes, or even the wherewithal to succeed. It only means that to protect yourself from further economic, emotional, and psychological trauma, you should get a regular job and make more solid career plans.

This may sound cold but it is important that you get on with your life and live it instead of just waiting for a script to sell. And it may be an ethical thing for you to do for yourself and your loved ones. We're talking here about the ethics of relationships as related to art.

People tend to make excuses for artists. They'll excuse crazy behavior, nutty appearance, even downright belligerence by saying "Oh, but Binky's a genius." That wears pretty thin over the years and genius or no, your friends and loved ones will leave you if you continue to be eccentric and egocentric without demonstrating that your work brings results.

So the bad news is: Know when to quit. The good news is: It will ultimately feel good to stop banging your forehead with a hammer. And if you don't want to quit altogether, change that hammer into a rubber mallet and write on weekends or after hours. At least you won't be thought of as a complete derelict or bum. And you won't think of yourself that way.

The universe does reward work—although not always in the order you think it should. For example, you may be rewarded for your writing by making money at Pigs-R-Us. We just have to break up the body–mind partnership that makes us believe that our income has to come directly from the thing we most love to do.

And you know that New Age maxim "do the thing you love and the money will follow"? I don't for a second buy it. People who keep touting that maxim have never spent years in the arts! People in the arts are not followed around by money. You can be followed around by money only if you're lucky enough to be entrepreneurial and love a profession that is easily integrated into the market economy. The trick is to figure out how your writing can make you a living in other ways (write advertising, do journalism, write PR releases or greeting cards) while you turn out the screenplays you love to write.

The problem with screenplays is that they don't really have a life if they aren't made. People don't sit around reading them for pleasure. They aren't considered works of art by themselves. So the screenwriter is trapped. One must be "produced" to communicate "one's vision." Otherwise, the work might just lie there like so much phosphorescence on the surface of the sea—going nowhere.

What a mess???!! It all sounds daunting doesn't it? So why do it? Why sit alone visualizing and writing down what you visualize if it's a thankless task and a hard one? Why take constant criticism, scorn, derision, rejection over and over again? It's a good question and one that Tolstoy asked in his wonderful little book *What is Art?* Tolstoy wrote it in 1898 and it's still amazingly modern and applicable. It's by a writer of prose (not a theologian or philosopher) who thought deeply about the purpose and the effect of his work. I'll also be using quotes from other writers who never wrote screenplays...writers who were deep thinkers and who created great literature.

Tolstoy wondered at a "profession" that consumes so much energy and, sometimes, even human life. He wondered at the punishing rigors of artistic professions, that people "kind and clever and capable of all sorts of useful labor, grow savage of their specialized and stupefying occupations and become one-sided and self-complacent specialists, dull to all the serious phenomena of life and skillful only at rapidly twisting their legs, their tongues, or their fingers." He wondered not only at what he calls the "stunting of human life" but also at the merciless psychological pressure and ruthless abuse meted out by critics, directors, even colleagues and friends.[7]

For whom, Tolstoy asked, is all this being done? Whom can it please? These are very good questions indeed. He asked if perhaps it is all being done for "art's sake." "Is it true," he asked, "that art is so important that such sacrifices should be made for its sake?"[8] I say

---

[7]Tolstoy, Lev Nikolaevich. *What is Art?* Oxford University Press, London, 1932, pp. 70–75.
[8]Ibid., p. 79.

no. The reason we do it all isn't for the sake of art, which, as Tolstoy said, is becoming something more and more vague and uncertain to human perception, its purpose more and more obscure. (The next chapter examines these concerns).

There are only two good reasons for writing and for enduring all that comes with that activity: Having Fun and Having Something to Say.

YUKS: AS GOOD AS BUCKS

Let's talk about the fun first. Keep reminding yourself, there's nothing wrong with having fun. It makes you happy and everyone craves happiness and works hard to get it. Books are written on happiness. Religions are founded on it. Some say it's what really gets us revved in life.

Screenwriting makes me happy because it's fun. Like backyard lounging or eating chocolate, it inspires and delights me. When I'm writing, time flies, there are no worries, and the only thing that's important is solving that puzzle on the page. For me, it's always been like being involved in a mysterious alchemy, a process as exciting to me as finding a cure for claustrophobia, or (even though the Leonard Cohen song says there is no cure) for love.

When I'm really writing I don't hear the music I have on in the background. I don't hear the telephone ring. I don't hear the nagging voices that tell me I'm crazy to be writing movies at all. The only thing on my mind is the work before me, and even though most of the time it's difficult—even maddeningly so—it's still the most exciting and thrilling thing there is. The creative act of writing is a physical, mental, and emotional high that no drug or drink can equal.

For me, writing has always been a life raft that's floated me through the most difficult times. It's always helped to pass the time during boredom and even made physically arduous activities bearable. For example, long before I knew about Solzhenitsyn, when I was in high school (my own idea of a gulag), I used to walk 2 miles to school. During those treks, especially in winter (in knee-deep snow just like the cliché), I'd compose short stories and memorize them. They kept my mind away from thoughts of cold and wet feet. Those stories paid off during composition exams, when I'd dredge them from my memory and copy them down to win As.

When I write, the worlds I visualize and the process of doing that seem so pleasurable that sometimes re-entry into "normal" life is painful. I think that's why some writers are drunks and drug addicts. Re-entry into the "real world," away from the one created from the energy of craft, is difficult. There is less control in the "real" world, less beauty, and far less apparent truth.

As distinguished South African novelist Nadine Gordimer said,
"The solitude of writing is quite frightening. It's quite close sometimes to madness, one just disappears for a day and loses touch.
The ordinary action of taking a dress down to the dry cleaners or
spraying some plants infected with greenfly is a very sane and good
thing to do. It brings one back, so to speak. It also brings the world
back."[9]

But even so, I still find it fun. I especially find screenwriting fun because
of the magic possibilities inherent in it. This became most clear to me
when I was sitting in a theater watching the film of my very first screenplay,
Home Free. I had written that screenplay about an incident from my own
childhood. Of course, I had made it dramatic and embellished it some but
it was still pretty accurate. And I had also included little details in the description of settings that were particularly important to me.

As the lights went down and the film started, I began to actually
see the events that heretofore I had only remembered. They came to
life again, made actual by light. Other people could see them exactly
as I saw them. It was an electrifying and almost spiritual experience. I had imagined something and then had given it life outside my
own mind! Since then, every act of screenwriting has been a magical
experience for me in the anticipation of imaginary or remembered
events becoming actualized in space and time. Screenwriting is
wound up in my love of movies and in my delight at the power they
have to make stories "real."

Of course, the minute stories are written down, they become essentially "real" if only in the mind of the writer. And this knowledge is
also a source of real pleasure to writers. Here's what some famous
writers have said about the pleasure they derive from writing.

Ernest Hemingway:

"It's more fun than anything else."

*"I have to write to be happy whether I get paid for it or not. But it is a
hell of a disease to be born with. I like to do it. Which is even worse.
That makes it from a disease into a vice. Then I want to do it better
than anybody has ever done it which makes it into an obsession."[10]*

---

[9]Plimpton, George (Ed.). *The Paris Review Interviews Writers At Work*,
Penguin Books, Sixth Series, 1985, p.275.
[10]Phillips, Larry (Ed.). *Ernest Hemingway on Writing*, Charles Scribner's
Sons, New York, 1984, pp. 14–16.

Ray Bradbury:

If you are writing without zest, without gusto, without love, without fun, you are only half a writer. It means you are so busy keeping one eye on the commercial market, or one ear peeled for the avant-garde coterie, that you are not being yourself. You don't even know yourself. For the first thing a writer should be is—excited. He should be a thing of fevers and enthusiasms. Without such vigor, he might as well be out picking peaches or digging ditches; God knows it'd be better for his health."[11]

Toni Morrison:

"(The work of art is) a haven, a place where it can happen, where you can react violently or sublimely, where it's all right to feel melancholy or frightened, or even to fail, or to be wrong, or to love somebody, or to wish something deeply, and not call it by some other name, not to be embarrassed by it. It's a place to feel profoundly."[12]

BECAUSE IT MATTERS

And that brings us to the second reason to write: Because you have a story you have to tell, because you have something important to discover about yourself and life; in other words, because you have something to say. This may be the most important reason. And strangely, it's also tied up with pleasure because, for some reason, saying what you believe or know to be true gives you a wonderful feeling indeed.

There's nothing more self-declarative than making a statement about how you see the world. I think it's an attempt to reach out to another person, shake that person by the brain, and make the purest kind of human contact—a transmission of thought. To me, telling someone what I think, in written form, is an act of brotherly love. It's my way of being human and I believe that communicating with language is one of the things humans were put on earth to do. It's my belief that if we can communicate in that way, and we are understood and understand, then we can achieve peace and harmony with mankind and turn up the light in the world.

Of course, you may not know exactly what you think about a thing until you begin to write about it. You don't necessarily have to. You

---

[11]Bradbury, Ray. *Zen in the Art of Writing*, Capra Press, Santa Barbara, 1989, p. 4.

[12]Ruas, Charles. *Conversations with American Writers*, Alfred A. Knopf, New York, 1985, p. 234.

may have an idea, a thought about life that you want to explore, and writing is a legitimate, even wonderful way of discovering your personal point of view. But in order to make that discovery, you first have to be willing to declare what you are seeking and to define the area where your search will begin.

This takes a commitment to a subject. It takes trusting that you do have something to say and that you do want to define and/or clarify it.

Bernard Malamud said "Very young writers who don't know themselves obviously don't know what they have to say. Sometimes by staying with it, they write themselves into a fairly rich vein."[13] And yet even young writers know there's something they want to discover about life and they want to share that discovery with an audience. Writing is something we have to do and good writing comes from a very deep place inside of us. Here's what, some famous writers say about that:

Norman Mailer:

"Jean Malaquais ... always had a terrible time writing. He once complained with great anguish about the unspeakable difficulties he was having with a novel. And I asked him, 'Why do you bother with it?'... He looked up in surprise and said, 'Oh but this is the only way one can ever find the truth. The only time I know that something is true is at the moment I discover it in the act of writing.' I think it's this moment when one knows it's true. One may not have written it well enough for others to know, but you're in love with the truth when you discover it at the point of a pencil. That, in and by itself, is one of the few rare pleasures in life."[14]

Anna Hamilton Phelan:

"I think that you just have to write from your heart, though many people and your friends and your family will tell you it's not commercial, but you just have to go ahead and write it if you want to write it. It has to come from your heart, because if you write it from your head or your wallet ... forget it."[15]

---

[13]Plimpton, George (Ed.). *The Paris Review Interviews Writers at Work.* Sixth Series. Penguin Books, New York, 1984, p. 157.

[14]Plimpton, George (Ed.). *The Paris Review Interviews Writers at Work.* Third Series. Penguin Books, New York, 1967, pp. 277–278.

[15]Froug, William. *The New Screenwriter Looks at the New Screenwriter,* Silman-James Press, Los Angeles, 1991, p. 31.

F. Scott Fitzgerald:

"You don't write because you want to say something, you write because you've got something to say."[16]

Steve Martin:

"And it's having something to say. In my case, I've lived long enough. Learning how to open those doors up, to understand you have something to say. It took me this long to finally discover I had something to speak about."[17]

Charles Burnett:

"I respond to things that are decent and character-driven, that open doors to things about human nature, reveal some insight into problems and make the characters have to think. Therefore, so does the audience."[18]

David Cronenberg:

"One of the things you want to do with any kind of art is to find out what you're thinking about, what is important to you, what disturbs you. Some people go to confession or talk to close friends on the phone to do the same thing."[19]

"Shakespeare is renowned for being all of his characters. They come to life because it's obvious that he could get right into their heads and understand them. I create characters I have to create from inside me."[20]

William Kelley:

"I think. Therefore I am. I write. Therefore I am next to God."[21]

---

[16]Fitzgerald, F. Scott. http://madscreenwriter.com/writequotes.htm

[17]Stayton, Richard, Steve Martin. *Written By Magazine*, June/July 1999, Vol 3, Issue 6, Los Angeles, p. 26.

[18]Harris, Erich Leon. *African American Screeenwriters Now*, Silman-James Press, Los Angeles, 1966, p. 27.

[19]Rodley, Chris (Ed.). *Cronenberg on Cronenberg*, Faber and Faber, London, 1993, p. 26.

[20]Beker, Marilyn. David Cronenberg, *Expression Magazine*, March/April, 1989, p.155.

[21]Kelley, William. 2001 Las Vegas Screenwriters Conference, *Lifetime Achievement Award Acceptance Speech*, July 15, 2001.

If, like the writers quoted, you take great pleasure in writing and have the urge to communicate something about what matters to you, you'll never get tired of trying to get your word out, of trying to connect. If those two things motivate you, then nothing else matters. You'll go on doing it because it's who you are and what you do.

And that will be especially important to you if you want to make a commitment to writing in the face of some very real economic threats and social negativity. It takes a lot to stick to writing (let alone screenwriting) when the rest of the world, your family, friends, and sometimes even the person you love the most, think it's a fly-by-night, impractical, and even foolish way to spend your time. But if your motivation comes from a genuine love of the craft, a belief in yourself, and a burning desire to say something, then nothing can stop you. Even if you have to take other full-time jobs and sneak your writing into a busy life, you'll do it and you'll find yourself satisfied and fulfilled.

To help you determine your motivation, try the following:

Exercise

1. Write down what motivates you to write screenplays. Take a couple of pages to do it if you need to. Be honest. Your motivators can be as wacky or weird as you want, as long as they are very important to you. Then synthesize your list into as few words as possible. Write them in large letters on a card and place them by your computer. Reread them when you start to write and when you feel frustrated, anxious, or blocked. Reassess them often and repeat the process when necessary.

2. Write down everything that scares you about writing. Include economic concerns, what other people think, and so on. Then try to synthesize this list into a few words. Write that down. What you write down is the identity of your mental enemy—the force that will keep fighting your motivation. Knowing what's stopping you (your enemy) is the first and maybe most important step in fighting to keep motivated and energetic about your screenplays. Then, take that "fear" list and tear it up. Burn it. Stomp on it. Flush it down the toilet. Determine that it will not be stronger than your motivation. Each time it comes up for you, focus on your motivation list. At least while you are writing, make your mind an impenetrable fortress against fear and doubt.

3. Get script specific. For each script, write down what you are trying to say in the script. Put this in one simple sentence if you can

(e.g., True love is worth all the trouble it takes). Consider what impact your thesis might have on an audience. Does it add to the well-being of humanity? Does it make the world a better place? Is it life-affirming? Will you be able to make your statement with a clear conscience? Stick this up somewhere on or near your computer screen and as you are writing consider whether or not each scene of your script somehow serves your thesis.

Now consider whether a "humane" message is important in a work of art. Shouldn't artists have the right to say *anything* no matter what its impact on society?

SOCIAL RESPONSIBILITY

Most people think they are socially responsible but they shy away from coming up with a workable definition of what that means. Watchdog media groups form to advocate social responsibility in film and television but even they won't be clear about what it means to be socially responsible. "I'll know it when I see it" isn't a satisfying answer.

Some people insist that social responsibility means not adding to or encouraging social disruption. But if you look at history, it quickly becomes obvious that in some societies, and in some eras, it was considered socially responsible to stand by and say nothing while people were being persecuted or killed, and even to engage in that kind of murder. If we take Nazi Germany as an example, we see many instances where citizens thought they were being "socially responsible" to an Aryan society by killing Jews or allowing them to be killed. In Rwanda, and the former Yugoslavia, "ethnic cleansing" took place in the name of a certain kind of "social responsibility." In fact, many crimes and oppressions (including racism in America) have taken place to keep order and reinforce a societal status quo.

It's important not to confuse responsibility to a specifically defined society with responsibility to the broad "society" of humanity as a whole. Too often, people consider "society" to mean political structure or social philosophy and are responsible (and urge others to be responsible) to the prevailing sentiments of the moment.

This kind of social responsibility changes with the times and can be very volatile and repressive. Again, looking at history, we can find examples of repressive social responsibility in the China of the Cultural Revolution, in Communist Russia, in Burma, and Chinese-controlled Tibet. The Chinese condemned the democratic protestors in Tiananmen Square as socially irresponsible. Russia did the same when its regime was most totalitarian. Myanmar's junta sanctioned Aung San Suu Kyi, who won a Nobel prize for her Burmese democratic work.

And then there are the "socially responsible" sin patrols of Malaysia—morality police who go around the country arresting, imprisoning, and fining people they say have transgressed the laws of Islam (kissing in public, drinking alcohol, gambling, committing adultery, practicing homosexuality, insulting Islam, or eating in public during the holy month of Ramadan). These people think they are keeping society safe and moral, as did the Taliban in Afghanistan, who, because of their beliefs, kept women from working and getting an education, as does Saudi Arabia, a country in which women are not allowed to drive.

To Americans, these kinds of "social responsibilities" are far-fetched and odious. And yet, Americans have to be very careful that we don't unconsciously, and even insidiously, make the same mistakes. We have to be careful that we protect the freedom that allows each one of us to express our own opinions, no matter how diverse or contrary to those in the majority.

At the same time, I believe that writers do have a responsibility to inform, inspire, and/or nurture the human race by at least striving to elevate its consciousness.

That's why I define social responsibility in a humanistic way by saying that to be socially responsible one does not engage in behavior or generate messages that will encourage, advocate, or otherwise contribute to inhumane behavior by others. By this definition, social responsibility means *not* taking part in, advocating, encouraging, or inspiring violence, unethical behavior, blatant inhumanity, or debasing acts that would add to the chaos, isolation, and brutality of modern culture.

Pope John Paul II wrote: "Man is free and therefore responsible. He has a personal and social responsibility, a responsibility before God, a responsibility which is his greatness."[22] And yet those involved in the business of creating entertainment argue that their chief "responsibility" is to accrue revenue for their companies and that "social responsibility" is the province of religion and public broadcasting, and not the entertainment industry.

That may be true for screenwriters if they view their work entirely as a business, but I believe that the majority of screenwriters do not look at their work that way. I believe screenwriters see themselves as writers, artists, and creative people first and as "workers" in an industry secondarily. I also believe that screenwriters' first allegiance must be to their own visions and not to the dictates of the marketplace.

---

[22]John Paul II, Vittorio Messori (Ed.). *Crossing the Threshold of Hope*, Alfred A. Knopf, New York, 1997, p.180.

But because there are no right answers here and no absolutes, each of us has his or her own definition of social responsibility. What's important is that each one of us examines what that is. Only then will we be able to really make a case for or against the need for social responsibility in creative work.

Exercise

1. Write down what you think social responsibility means. Be specific. "Not hurting anyone" is too general. Try to hone what it means to you. You'll find that this is difficult to do and that what you write may surprise you. You may also notice that your "definition" might want to change with the circumstances. What are the circumstances? Public opinion? Laws? Governments? Cultural climate? Think about how social responsibility changes with the times. Consider your part in this.

2. Consider your role in "society." By that I mean, think about yourself as an individual within a culture and consider what that means. Those of us who live in a democracy should have greater responsibility within that democracy. In the United States we are empowered by the Constitution to express our opinion. It is not a luxury many have in foreign lands. And yet those in foreign lands who do not have democracy often are far more "revolutionary" and cognizant of their place in their culture than are Americans.

3. Write about your place in the culture and your society. Dissect this in narrow terms. Consider your place in your family, your household, your neighborhood, your work place, your town or city, your state ... and so on. Remember to make significant your work as a screenwriter. Then list the ways in which you might be "socially responsible" in each category and define what that means to you.

WHAT'S ART GOT TO DO WITH IT?

Heaven forbid that I should try to answer the question "What is Art?" Great writers, thinkers, jurists, and fools have tried to come up with a definition of "Art" for centuries and with little success. I'm not even going to try. We all have our own ideas of what Art—I'll call it *True Art* from now on—is. I'll tell you what True Art is for me.

As far as I'm concerned, True Art is usually something thoughtfully rendered, personal, provoking, surprising, unusual, original, creative, courageous, and independent. By that I mean, that True Art makes a statement so strong and transmits a vision so powerful in its originality that it skirts and sometimes even flaunts convention.

And, most important, True Art packs an emotional, intellectual, and even visceral wallop. Does True Art need to be socially responsible? I would hope that those practicing it would think so, but it may not matter. True Art exists almost outside of society even though it may enlighten and benefit society. True Art is not dependent on audience or audience reaction. True Art may be created for its own sake, for the artist's sanity or obsession, or for its own small audience, and it is satisfied with that.

These ideas have been shaped by the conditions of my life, my education, and my environment. My early life, for instance, had a great bearing on how I define True Art.

My parents were European immigrants who revered education and encouraged reading. The public library was a sacred place for me. By the time I was 10, I had read all the books in the kids' section and as I made my eager way upstairs into the adult section (graduated by special dispensation of an understanding librarian) I felt like Heinrich Schleiman discovering Troy. So overwhelmed was I by the treasures that I found in that adult library, I determined to challenge myself by searching out obscure literature rather than the popular kind. This developed in me a keen interest in story and the elegance

47

of language, and established a respect for those original writers who strayed from the norm.

We didn't have TV in my house until I was about 10. TVs were big and expensive and we didn't have much money. I made do with books and with radio shows that were based on books. When I finally did watch TV, I gravitated to informational programs and documentaries because I had already determined that learning was the best entertainment possible. It helped that I grew up in Canada where much of TV and radio were publicly owned and documentary heavy.

The odd program we got shipped in from the United States through Buffalo (I stared for hours at the Indian-in-a-Headdress test pattern) seemed exotic: *Ed Sullivan*, *Gunsmoke*, *Perry Mason*. But those shows were for adults and saved for weekends when the whole family would watch them together. TV was entertaining and interesting but certainly it wasn't True Art—not when you compared it to Dickens, Austen, Brontë, and Swift.

Movies weren't True Art either. The first movie I remember seeing was a scratchy revival of *The Wizard of Oz*. The moment the screen changed from black and white to color I was hooked. From then on, I'd spend every Saturday afternoon at the Adelphi theater—a neighborhood rerun house specializing in double feature matinees for kids.

The old couple who owned the theater brought in films they thought "suitable for children" that they could get for almost nothing—B pictures from the 30s, 40s, and 50s (often in black and white). Twenty-five cents got you a small bag of popcorn and a seat in an packed rambunctious fleabag. Just before showtime, the theater lady (who always dressed in a white lab coat and hairnet: Cafeteria Matron meets Nurse Ratched) would line us up and collect our quarters. While we stood shivering or sweltering on the sidewalk (depending on the season), she'd bark out orders commando style. "No screaming in the theater. No fighting. No throwing stuff!"

As she'd stroll up and down the line staring at us and trying to weed out the troublemakers, I got the message that movies were important, and you had to earn the right to see them. But never for a moment did I consider them True Art because the audience I was a part of was far too scruffy and irreverent to appreciate True anything. That was evidenced by our behavior once inside the theater where all bets were off.

During screen fistfights, real fights would break out in the aisles. During kissing scenes, we booed and threw things. During the boring bits, we jumped up and down on the seats or crawled on the floor between them (we didn't worry about floor gook then!) and pretended to be spies, cowboys, or robbers. If the dialogue was bad, we'd make it up, shouting back at the screen with vigor. It was all splendid fun but it wasn't True Art. That took place in an arena of reverence and silence with no popcorn and no fights.

As I grew older, and particularly after I started to care about dialogue, I did become more reverential about movies. Talking in movies was anathema. (At the Writer's Guild theater you get thrown out for talking, and rightly so.) Even cheering or sobbing loudly was verboten. Movies to those of us who took them seriously enough to have careers working on them became all-consuming and special. But not everyone thinks like that.

The reality is that most people go to movies to have fun. They still like to laugh loudly, make comments at the screen, and cheer when they feel like it. That doesn't mean that rowdy audiences are a good thing. (I detest them!) But it does mean that movies are made to be enjoyed by a great number of people in an atmosphere where eating, whispering, and inattention takes place, and to me that doesn't make them True Art. So, by my definition, precious few movies are True Art.

True Art films are usually small, independent, and obscure. They have a long shelf life but very little public life. They are not generally consumed by the masses. Films like *Le Chien Andalou*, early stuff by Warhol, Michael Snow, other experimental filmmakers, strange foreign offerings, have very small though loyal audiences who seek them out.

Movies (even "artistic" ones) that are consumed by the masses are Popular Art. Popular Art creates, records, and even glorifies popular trends. It speaks to popular concerns and whims. Movies are Popular Art because they are dependent on popular tastes and cater to and even inflame the desires of the mass market . Movies are considered a "product" by an "industry." Movies are expensive to make. They usually don't get made without careful thought to what audiences will pay to see. Movies are usually audience-tested before completion and then changed according to market research. Movies don't hold much stock in personal vision. They may start off that way but because of their collaborative and product-to-market nature, they rarely are personal reflections of a single artist. Certainly they almost never are the personal reflections of a screenwriter unless of course that writer also directs, produces, and pays for the film. And even then that person is at the mercy of the distributor who might want to call the shots.

This doesn't mean that there aren't exceptions. There are some filmmakers who go on making personal films that do get funding, distribution, and audiences (John Sayles, Jim Jarmush, and Alison Anders are examples) but those film makers are rare and work on the fringe. I also don't mean that popular films can't have artistic qualities. Of course they can. All popular films in fact are "artistic" by virtue of their mise en scène and their ability to conjure up actual worlds.

But unlike True Art films, movies in general or even limited release have a profound effect on a great number of people. So while I think that True Art doesn't necessarily have to be "socially responsible" because it reaches so few people and must find its own audience, I do think that movies should be. Too many studies have shown that popular images affect people, coerce them to buy things and think things and even do things they might otherwise not buy, think, or do.

Indeed, images wield so much power that lives—even whole societies—can be forever changed by them. The advertising industry (and some would say our whole consumer economy) exists based on this notion. Of course, a work of True Art has the same power but not on such a large scale. The reality is that all Art-True and Popular, is a kind of propaganda for the ideas and ideals of the artist. It's especially important then for the artists who work in popular art to be conscious of the messages they proffer so they can wield their power responsibly.

Does True Art need to be in service of humankind? Not necessarily. Does Popular Art? I think it should. And I think that way because my father inspired me to. As I already told you, my father was an immigrant. He came to North America when he was 35 years old. He didn't speak a word of English and had no real skills but he was very smart, brave, exceptionally good-looking, had a magnetic personality, and a sensational sense of humor. He eventually worked his way up to owning and operating a small slipper factory. He and his partner did all the sewing and shipping themselves. My father also made all the sales calls.

Eventually, after years of backbreaking hands-on work, he was able to hire a small crew of workers to help him. He hired people whom he thought needed a break and who were immigrants like himself. Most of them were Italian and couldn't speak a word of English so my father learned some Italian. He treated his employees like friends and worked right alongside them. He built such a rapport with them that many stayed with him for over 20 years.

My father didn't like the factory work much. He loved selling but he hated sewing. He did it anyway. And when I asked him why he didn't quit the business his answer was one that continues to inspire me to this day. He told me that his workers depended on him for their jobs and that it made him feel good to know he made a quality product that people would enjoy using. The true worth of any occupation, he told me, was its service to others.

If that's true of slippers, then, in my opinion it should certainly be true of popular art. So although artists may deal with debasing acts, blatant inhumanity, or violent and unethical behavior in works of art, the degree to which those are depicted in order to make a point should be ameliorated by the size and scope of the audience for which the works are intended. The audience should always be respected and considerately served. Just because movies are artistic and screen-writers call themselves artists, doesn't mean that anything goes.

PART II
THE CERTAINTY OF WHAT: ANYTHING GOES?

A GLIMPSE OF STOCKING

"Days of yore a glimpse of stocking
was looked on as something shocking
today heaven knows, anything goes."

—Cole Porter (1934)

hen Cole Porter wrote the lyrics for his song "Anything Goes," public "morality" as it pertained to sex was far different than it is today. And rightly so. I don't think any of us want to go back to the days when women fainted if men saw their ankles or were labeled loose if they crossed their legs. And certainly no one wants to go back to the days of the Hays Code.

The Hays Production Code, named after its first director, Will Hays, was a "voluntary" code formulated and adopted in 1934 by The Association of Motion Picture Producers and the Motion Picture Producers and Distributors of America to "protect" public morality by controlling (and censoring) movie content. Highly restrictive and repressive, it outlined in detail how (and if) particular subjects should be presented to audiences. Will Hays and his appointed censors were the final authority determining code interpretation and application. No movie could be distributed in America without the Hays "seal of approval."

Because everyone who condemns censorship always talks about how horrible the Production Code was (and rightly so), I'm including its general principles and particular applications here. If you want to read the entire code, including the preamble and reasons for support-

ing that preamble, you can look it up on the web at www.
artsformation.com. As you read them, you can see how dangerously
vague some of these "rules" were and how much power was wielded by
the censors whose job it was to interpret and enforce them.

The Motion Picture Production Code (Hays Code)

General Principles

1. No picture shall be produced that will lower the moral standards
   of those who see it. Hence the sympathy of the audience would
   never be thrown to the side of crime, wrongdoing, evil, or sin.
2. Correct standards of life, subject only to the requirements of
   drama and entertainment, shall be presented.
3. Law, natural or human, shall not be ridiculed, nor shall sympathy be created for its violation.

Particular Applications

I.   Crimes Against the Law.
These shall never be presented in such a way as to throw sympathy
with the crimes as against law and justice to inspire others with a desire for imitation.

   1. Murder
      a. The technique of murder must be presented in a way that
          will not inspire imitation.
      b. Brutal killings are not to be presented in detail.
      c. Revenge in modern times shall not be justified.
   2. Methods of Crime should not be explicitly presented.
      a. Theft robbery, safe-cracking and dynamiting of trains,
          mines, buildings, etc., should not be detailed in method.
      b. Arson must be subject to the same safeguards.
      c. The use of firearms should be restricted to the essentials.
      d. Methods of smuggling should not be presented.
   3. Illegal drug traffic must never be presented.
   4. The use of liquor in American life, when not required by the plot
      or for proper characterization, will not be shown.

II. Sex

The sanctity of the institution of marriage and the home shall be up-
held. Pictures shall not infer that low forms of sex relationship are the
accepted or common thing.

1. Adultery, sometimes necessary plot material, must not be explicitly treated or justified, or presented attractively.
2. Scenes of Passion
   a. They should not be introduced when not essential to the plot.
   b. Excessive and lustful kissing, lustful embraces, suggestive
      postures and gestures are not to be shown.
   c. In general passion should so be treated that these scenes
      do not stimulate the lower and baser element.
3. Seduction or Rape
   a. They should never be more than suggested, and only when
      essential for the plot, and even then never shown by ex-
      plicit method.
   b. They are never the proper subject for comedy.
4. Sex perversion or any inference to it is forbidden.
5. White slavery shall not be treated.
6. Miscegenation (sex relationships between the white and black
   races) is forbidden.
7. Sex hygiene and venereal diseases are not subjects for motion
   pictures.
8. Scenes of actual child birth, in fact or in silhouette, are never to be
   presented.
9. Children's sex organs are never to be exposed.

III.   Vulgarity
The treatment of low, disgusting, unpleasant, though not necessarily
evil, subjects should always be subject to the dictates of good taste and
a regard for the sensibilities of the audience.

IV.   Obscenity
Obscenity in word, gesture, reference, song, joke, or by suggestion
(even when likely to be understood only by part of the audience) is
forbidden.

V.   Profanity

Pointed profanity (this includes the words, God, Lord, Jesus, Christ—unless used reverently—Hell, S.O.B., damn, Gawd), or every other profane or vulgar expression however used, is forbidden.

VI. Costume
1. Complete nudity is never permitted. This includes nudity in fact or in silhouette, or any lecherous or licentious notice thereof by other characters in the picture.
2. Undressing scenes should be avoided, and never used save where essential to the plot.
3. Indecent or undue exposure is forbidden.
4. Dancing or costumes intended to permit undue exposure or indecent movements in the dance are forbidden.

VII. Dances
1. Dances suggesting or representing sexual actions or indecent passions are forbidden.
2. Dances which emphasize indecent movements are to be regarded as obscene.

VIII. Religion
1. No film or episode may throw ridicule on any religious faith.
2. Ministers of religion in their character as ministers of religion should not be used as comic characters or as villains.
3. Ceremonies of any definite religion should be carefully and respectfully handled.

IX. Location
The treatment of bedrooms must be governed by good taste and delicacy.

X. National Feelings
1. The use of the Flag shall be consistently respectful.
2. The history, institutions, prominent people and citizenry of other nations shall be represented fairly.

XI. Titles
Salacious, indecent or obscene titles shall not be used.

XII. Repellent Subjects
The following subjects must be treated with the careful limits of good taste:

1. Actual hangings or electrocutions as legal punishments for crime.
2. Third degree methods.
3. Brutality and possible gruesomeness.
4. Branding of people or animals.
5. Apparent cruelty to children or animals.
6. The sale of women, or a woman selling her virtue.
7. Surgical operations.

The Code was problematic because it dictated content and censored according to standards that were defined, determined, and enforced by a body of its own choosing and it wielded ultimate power. As odious as this was, the Code was in place (albeit loosely enforced after 1960) until 1966 when the new rating system emerged!

The Rating System

Inducement for a new system partly came from the climate of the late 1960s. As Jack Valenti, president of the Motion Picture Association of America put it:

> The national scene was marked by insurrection on the campus, riots in the streets, rise in women's liberation, protest of the young, doubts about the institution of marriage, abandonment of old guiding slogans and the crumbling of social traditions. It would have been foolish to believe that movies, that most creative of art forms, could have remained unaffected by the change and torment in our society.[23]

Mr. Valenti said that the first film he scrutinized as new head of the MPAA was *Who's Afraid of Virginia Woolf*, in which for the first time on the screen the word "screw" and the phrase "hump the hostess" were heard. It was also the first time a major distributor was marketing a film with nudity in it. Valenti backed the decision of the Production Code Administration in California to deny the seal of approval to the film.

Undeterred, MGM distributed the film through a subsidiary company, thereby flouting the voluntary agreement of MPAA member companies that none would distribute a film without a Code seal. The breakdown of that mutual agreement was the beginning of the end for the Hays Code. And it meant that films would and could ultimately be shown as the filmmakers intended—without censorship.

---

[23]Valenti, Jack. How It All Began. www.mpaa.org/movieratings

In the face of the "revolution" by distribution companies, on November 1, 1968, the MPAA announced a new voluntary film rating system with specific categories in place that would guide the public in discerning the nature of film content. This system was put in place especially to "protect" children, in keeping with the U.S. Supreme Court decision in April, 1968 to uphold the constitutional power of the states and cities to prevent the exposure of children to books and films that could not be denied to adults.[24]

These categories have been adjusted over the years, but for the most part they are still in play today. They are listed here, including, where applicable, a brief description of the adjustments made over the years, to each category.

G for General Audiences, all ages admitted

M for mature audiences—parental guidance suggested, but all ages admitted. Soon though, this category was adjusted to GP meaning General audiences, Parental guidance suggested, and then PG: Parental Guidance Suggested. On July 1, 1984, the PG category was split into two groups, PG and PG-13. PG-13 meant a higher level of intensity than was to be found in a film rated PG.

R for Restricted. Children under 16 would not be admitted without an accompanying parent or adult guardian (later raised to under 17 and varies in some jurisdictions). On September 27, 1990 two more revisions were introduced, brief explanations of why a particular film received its R rating. Sometime later the explanations were applied in the PG and PG-13 ratings.

X for no one under 17 admitted. Anyone not submitting his or her film for rating could self apply the X. The name of the X category was changed to NC-17: No one 17 and under admitted and explanations were applied.

How are these ratings decided? A full description of that process is given on the MPAA website. Briefly though, labeling is decided by a full-time Rating Board located in Los Angeles and made up of 8 to 13 members. Valenti wrote that "members must have a shared parenthood experience, must be possessed of an intelligent maturity and most of all have the capacity to put themselves in the role of most American parents so they can view a film and apply a rating that most parents would find suitable and helpful in aiding their decisions about their children's movie going."[25]

---

[24]Ibid.
[25]Ibid.

The rating system also works to inform adults about a film's content. That's a good idea. It makes sense to find out beforehand if a film contains violence, nudity, or strong language so that you can exercise your freedom to monitor what images you personally want to subject yourself to. There's nothing worse than sitting in a theater and being unpleasantly surprised by something entirely disagreeable.

Some may argue that this is the exact purpose of art—to surprise and shock the system. That might be so, but certainly consumers of that art should have the right to decide whether they want to expose themselves to that surprise and shock. Because of the compelling nature of film images and their profound effect on the psyche, there are certain things that people might absolutely NOT want to expose themselves to. That's a personal choice that people who pay money for a product have the right to make.

Personally, I simply can't stomach the sight of animals being hurt. I want to know if there are scenes of brutality against animals in a film before I see it so I can decide for myself if I want to endure the experience of those images in order to see a film.

Where adults are concerned, the Rating System is nothing more than an informational tool. Problems arise though, when parents expect that the Rating System will be used by theater owners to police the viewing habits of children and teenagers. Most people know that theater owners seldom, if ever, enforce a PG, PG-13, or R rating. Theater owners want to make money so their approach to ratings enforcement is entirely lackadaisical and although that approach may be borderline unethical, it's certainly understandable. But what excuses do parents have for not monitoring what their children see? I've seen lots of teenagers and even young children at films with R ratings. And some of them have been accompanied by parents. This kind of "parental guidance" is extremely ill-advised and patently unfair to other audience members who would rather not see graphic films in the company of youngsters.

In television, the FCC still monitors the public airways closely and networks exercise Standards and Practices that don't allow certain types of images and language on television that can be seen by a general public. For people who don't want to be restricted from watching what they want to watch on television, cable has provided a workable alternative.

Ultimately, most of us are exceptionally glad that we can see any kind of films that we choose to see. We're glad that, at least where movies are concerned, we're out from under organizational "parental guidance."

But if audiences are overjoyed at the viewing freedom afforded them by the absence of restrictions, filmmakers are especially ecstatic. It does seem as if filmmakers can go as far as they want to go to express anything they want to express. Think of the infamous Sharon Stone's leg-crossing shot in *Basic Instinct* (1992). Director Paul Verhoeven and screenwriter Joe Eszterhas might have wanted to make the movie as "erotic" as possible, but be honest—how many of us really needed to glimpse Stone's privates in order to know that her character uses feminine wiles to get her way?

I'm certainly not defending the odious Hays Code or any other kind of censorship. The old mores that never would have considered that kind of shot were too restricting and hypocritical but in their own quirky way they did give rise to more creative screen plays. For example, the cute and titillating (for those times) device of "the Walls of Jericho" in that Frank Capra gem, *It Happened One Night* (1934, written by Robert Riskin), wouldn't have been there if it weren't for the censors fussing about a man and a woman spending the night in the same room.

In case you don't remember the story, here it is. A runaway heiress brat (Elie Andrews, played by Claudette Colbert) is intent on marrying a man her father doesn't approve of. He locks her up to prevent it. She escapes without a dime and pawns her watch to get bus fare to return to her intended.

On the bus, she meets Peter Warren (played by Clark Gable), a wily and wild reporter who discovers who she is and offers to help her in return for an exclusive story. Short on cash, Elie and Peter have to share a motel room. To ensure modesty and decorum (and to satisfy Hays Code rules II, III, IV, VI, and IX), Peter gallantly strings a rope between the *two single beds*, and throws a blanket over it to afford Elie privacy. He calls the barrier "the walls of Jericho" and its existence adds greatly to the sexual tension between the couple. We can see Peter smirking as Elie undresses behind the "wall" but our only indication of her progressive "nudity" is articles of clothing being flung over the top of the barrier (satisfying Hays Code Rule VI).

Imagine if that movie were remade today. Peter and Elie would probably have extended, graphic sex in the motel's double bed and then spend the rest of the movie bickering and boffing. Less interesting and far less "cute." Maybe we don't want cute anymore but we sure could use something more interesting than today's in-your-face gratuitous genital aerobics that seldom add much to the story except palm sweat.

Of course, what used to make money for studios, and still does, is titillation. People went to the movies in the old days to see what they couldn't see respectably in public—Colbert's thighs, Lana Turner's heaving bosom, passionate kissing, anything that might sneak by the censors. But these days, that's tame stuff. In fact, our movies come far closer to what I'll call "Adult Films" (and many call Pornography) than ever before. We're down to showing practically all of sex except for the looped close-ups of "ins and outs."

In fact, because of competition from the Internet and cable TV, more and more films (particularly in the foreign market) are blurring the line between mainstream and "adult" films by depicting graphic sex. As Kristin Hohenadel wrote in a *New York Times* article on July 1, 2001:

> *Danish director Lars von Trier's Idiots features actual penetration as does French director Bruno Dumont's Vie de Jesus and the French Director Catherine Breillat's Romance; brightly lighted plainly filmed fellatio between ordinary middle-aged actors is featured in French director Patrice Chereau's English language film that won the top award at the 2001 Berlin Film Festival (and will probably open in the U.S.); and hardcore, unsimulated sex performed by the porn actresses in the unrated French film Baise-Moi which opened in New York on July 6, 2001. This film was also characterized by its raw depiction of violence. (Interestingly, Baise-Moi was banned after just 3 days on Paris screens not only for its graphic unsimulated sex but for its violence.)*

These films are made mostly in Europe where none of the American sex taboos have an effect. (In fact, Basic Instinct was released in Europe in a version that is more sexually explicit![26]) Most filmgoers of a certain age remember sneaking into Swedish director Vilgot Sjoman's 1969 film *I Am Curious Yellow* (it was first seized by customs when it was shipped into North America), to catch brief glimpses of full frontal male and female nudity.

Nowadays, that's nothing. We've seen Ralph Fiennes's backside (in several movies), Harvey Keitel's front side (in *The Piano*) and the bare breasts of practically every famous and bit actress there is. Even Tom Cruise and Nicole Kidman (the once-married couple) "did it" for the camera on a closed set in *Eyes Wide Shut* and made it look real.

Soon, even in America, all that may differentiate "Adult Films" from regular movies might be the budget.

> *Killing Me Softly by Chinese director Chen Kaige (Farewell My Concubine), an erotic thriller starring Heather Graham and Joseph Fiennes that explores the precarious relationship between sex and love, al-*

---

[26]Maltin, Leonard (Ed.). *Movie and Video Guide*. Signet. Penguin Group, New York, 1997, p. 85.

*ready threatens to push the envelope. Unlike the actors in many Holly-
wood movies, Ms. Graham and Mr. Fiennes were completely naked
during the sex scenes—some of which involve bondage and sadomas-
ochism—and used no body doubles. The director said that while the
actors did not have sexual intercourse on film, he tried not to give too
much specific direction during the filming of the sex scenes "to see if
they could create that chemistry themselves".[27]*

If morality has shifted and the public presentation of explicit sex
isn't considered by the majority to be immoral anymore, it may be
unethical because its inclusion in more and more films makes it
hard on actors who might not want to bare all for the camera and who
resent the body-double–porn-star substitution. Graphic sex also
tends to exploit women and treats them as sex objects. And, more
and more, it's doing the same to men. Sharon Stone became a star
because of that risqué shot in *Basic Instinct*, but the similar exposure didn't really help Demi Moore in *Strip Tease* and in *Showgirls*,
it positively killed Elizabeth Berkley's career. Berkley was blamed
for the film's failure but that failure was the fault of the Paul
Verhoeven–Joe Eszterhas combination that banked on sleazy exploitation bringing home the bacon.

Ultimately, audiences stayed home, making the statement that perhaps they needed more than blatant sex and tawdry thrills to be entertained. I bet that most people, even though they might not want to
admit it, don't really want to be subjected to watching other people "get
it off" as the British say, or bare all in public. It might well be human
nature to want to see pretty people having sex in public (certainly
Larry Flynt, publisher of *Hustler Magazine*, tells us that is so), but the
majority of us probably find it embarrassing, uncomfortable, or too
distracting from the story.

Long scenes of blatant sex often detract from the momentum and
the impact of the story line. Certain films do depend on sexual detail
for characterization (*Last Tango in Paris, Looking for Mr. Goodbar,
Henry and June, Boogie Nights, Eyes Wide Shut*, etc.) but the majority do not. Lots of times, sex scenes are extended or included because
they'll add to box office appeal. A particularly glaring famous example? Melanie Griffith's nude vacuuming shot in Mike Nichols's
Working Girl (1988). That scene was laughably gratuitous.

---

[27]Hohenadel, Kristin. Film Goes all the Way (In the Name of Art), *New York
Times*, July 1, 2001, ARpp. 1 &20.

So is it ethical to write scenes of graphic sex and nudity? Not if these scenes are peripheral to your screenplay's message, if they exploit the natural prurient interest of audiences, and are simply a marketing and sales tool. And even if a screenwriter can justify his own interest in graphic sex and make a case for the message it puts forward, writing those scenes might not be ethical because of the negative nature of the message transmitted by them.

We can use Joe Eszterhas himself as an example of what I mean. On August 13, 2002, I was in Toronto reading the opinion page of *The Toronto Star*. My mouth fell open when I saw a hard-hitting piece by Eszterhas himself decrying smoking! Here's a brief excerpt from that piece:

> I've written 14 movies. My characters smoke in many of them, and they look cool and glamorous doing it. Smoking was an integral part of many of my screenplays because I was a militant smoker. It was part of a bad-boy image I'd cultivated for a long time—smoking, drinking, partying, rock'n' roll.
>
> Smoking, I once believed, was every person's right. Efforts to stop it were politically correct, a Big Brother assault on personal freedoms. Second-hand smoke was a non-existent problem invented by professional do-gooders. I put all these views into my scripts.
>
> In one of my movies, Basic Instinct, smoking is part of a sexual subtext. Sharon Stone's character smokes; Michael Douglas' is trying to quit. She seduces him with literal and figurative smoke that she blows into his face. In the movie's most famous and controversial scene, she even has a cigarette in her hand.
>
> I'm sure the tobacco companies loved Basic Instinct. One of them even launched a brand of "Basic" cigarettes not long after the movie became a worldwide hit, perhaps inspired by my cigarette-friendly work. My movie made a lot of money; so did their new cigarette.
>
> Remembering all this, I find it hard to forgive myself. I have been an accomplice to the murder of untold numbers of human beings.
>
> I am admitting this only because I have made a deal with God. Spare me, I said and I will try to stop others from committing the same crimes I did. Eighteen months ago I was diagnosed with throat cancer ..."[28]

---

[28]Eszterhas, Joe. *The Toronto Star*, August 13, 2002, p. A21.

Way to go Joe! Those of us who have always believed cigarettes were killers might well applaud his change of heart and his courage to go public about it. It's unfortunate, though, that he had to get cancer to finally "see the light." One wonders what it will take for him to admit that perhaps his portrayal of women might also be damaging and irresponsible.

Is it socially responsible to include cigarette smoking in movies? Knowing what we do about the evils of tobacco today, I agree with Eszterhas that it isn't. But how about including sexually graphic scenes in screenplays? To decide, you might want to think about how American culture popularizes sex to create consumers.

Popular culture's images are blatantly sexual. And whereas sex sells, is interesting to a great many people, and can even be called "art" in certain instances, doesn't its graphic and ubiquitous presence in our popular culture somehow deliver a message we may not want to perpetuate?

Although it may be a good thing that we Americans have been freed up enough to enjoy our bodies and not be as ashamed of their natural functions as we once were, perhaps we've become too blasé about these functions and too quick to promote them. Is it okay with you that 9- and 10-year-olds are dressing like hookers because of the fashion statements made by TV, film, and music stars brought on by our increasing acceptance of public nudity? Are you happy that teenagers feel pressured to engage in sex because they are constantly bombarded with messages that it's socially acceptable to be overtly sexual and permissive? Are you happy that it's no longer an anomaly to be a single mother or that teenage pregnancies are escalating at an alarming rate perhaps because casual sex is often depicted as the public norm?

A recent statistic in *The Los Angeles Times* (August 20, 2001) said that "The number of unmarried-partner households shot up 72% nationwide. The largest gains came in the Bible Belt and across the Great Plains." Another statistic in *The Los Angeles Times* says that 32.8% of all births in the United States are now registered as "non-marital."

These trends are alarming because of what they tell us about our society. The figures seem to indicate that increasingly, people are less inclined to make moral commitments, to honor personal responsibilities, to respect themselves and others, and to curb their appetites. This makes for unstable relationships and family breakups that have profound effects on children.

And, too often, people are disillusioned by organized religion and orthodox restrictions and demonstrate this disillusionment by engaging in behavior that is counterproductive to good physical and mental health.

You need to ask yourself if you want to live in a society becoming increasingly more fragmented, fast, and loose. If you are unhappy with the way things are going and would like to effect a change, you must take action, especially if you work in popular media, one of the most powerful influences on people's attitudes and lives.

Exercise

Honesty is a requirement of this exercise.

1. Write down exactly the kind and amount of sex you personally are comfortable with writing about and with seeing in the company of others. Don't worry, no one has to read this. You can be as explicit as you like. Just remember that this entails *COMFORT*—what you would be comfortable writing and what you would be comfortable having people see with *YOU* in the audience.

Consider how comfortable you would be reading the sex scenes you've written in front of actors and/or producers. Consider how comfortable you would be presenting this same material to your family and close friends—people you care about. Include all age ranges. Keep in mind that even young teenagers are sometimes allowed into theaters even though the film has an R rating.

2. When you've determined your comfort level, write a "sex scene" for a script you are writing or intend to write.

3. Then, answer these questions:
   - Is the graphic degree of the scene absolutely necessary to the film's thesis and the depiction of the character?
   - Will the character's portrayal or the film's thesis be diminished by the reduction or exclusion of the graphic sexual nature of the scene?
   - What message does the scene deliver to the audience about the characters and about the film's thesis and the writer's philosophy?
   - Does that message coincide with your own personal "morality" and with the effect that you want to have on society? Can you take responsibility for having written it with a clear conscience?

4. Now, just for fun, no matter what you answered or how graphic (or nongraphic) your scene was, challenge yourself by rewriting the scene so that it remains erotic and suggestive without being at all sexually graphic. Look at some old movies, for examples, an old Bogart movie where Bacall lights a cigarette and then hands it

to him. Let yourself be creative. Have fun. And avoid cheap cutaways à la Naked Gun: rockets going off, firecrackers exploding, waves dashing against the shore.

No matter what you've written in your sex scene, if you are true to your thesis and to yourself, if you did not write the scene with the intention of exploiting audiences' prurient interests, if you took into account the impact of sex scenes on audiences, considered the message the scene will deliver, believe that it is socially responsible and are willing to take responsibility for that message, you will have written ethically.

Some viewers might still say that your movie is immoral based on their own definitions of what constitutes morality but you can't help that. In the spirit of freedom of expression, the movie you've written will be your version of truth and you'll have written with a conscience.

Remember though, that it's a cop-out to be outrageously graphic in a movie and then, at its end, to punish those who performed the graphic (and perhaps immoral) acts. This will not make your movie any more "moral" or ethical. The end does not justify the means. Are the effects of watching 2 hours of graphic and depraved acts negated when these acts are punished at the end of the movie?

Make no mistake—the powerful effects of images are long-lasting even though the "moral of the story" may be that immoral acts are not profitable and that immoral people lose. Images of bad guys suffering tends to fade. A much smaller amount of time is spent seeing that suffering than is spent seeing the immoral acts.

Clearly immorality exists in the world and it's the screenwriter's job to take it on if that's what the story calls for. Problems arise when writers revel in immorality for the sake of titillation and to please the studio who needs to sell the product. Calling graphic material "art" doesn't make it so.

As Tolstoy said "A real work of art can only arise in the soul of an artist occasionally as the fruit of the life he has lived. But counterfeit art is produced by artisans and handicraftsmen continually, if only consumers can be found. The cause of the production of real art is the artists's inner need to express a feeling that has accumulated. The cause of counterfeit art is gain."[29]

---

[29]Tolstoy, Lev Nikolaevich. *What is Art?* Oxford University Press, London, 1932, pp. 266–267.

SOMETHING SHOCKING

Perhaps it's understandable when writers claim that graphic sex is art. The study of art history can support some of these claims. But it's a much tougher sell when they try to maintain that depictions of graphic violence can be considered art. Writers and studios use violence the same way they use sex—to titillate and sell. In fact, an entire genre specializes in violence and it's called Action–Adventure. Granted, some of the violence may be simply implied or representational (as in the James Bond movies) or cartoon-like "Sock and Pow"s (like the antics of Batman and Jackie Chan) but it is violence nonetheless. And unfortunately, the latest examples of that genre, like *The Fast and the Furious* (2001)—condemned by critics as a highly irresponsible approach to driving practices—are pushing the envelope of "tolerable entertainment violence" and moving into a much bloodier graphic, gratuitous (and so, unethical) realm.

Just like sex, the degree of violence people will tolerate varies. It's a sad fact, but what some people consider too violent, others think is tame. Morbid curiosity married to journalism's "need to know" philosophy has brought before us all manner of gruesome displays in our news and even in ad campaigns (the famous Benneton billboards that depicted graphic violence and seemed to "iconize" convicted murderers). People are even buying "art" by serial killers and bidding for their artifacts on eBay! We've become so desensitized that the fact of death and mayhem, or the sight of it, no longer seems to have the power to stir us that it once did. In fact, it takes considerably more gore than it used to, to get even the most squeamish of us to react.

This is a function of our detachment from our own relationship to life. Many of us have come to see life, other than our own, as something trivial and inconsequential. Its fragility is brought home to us again and again in all our media in such force that we are no longer surprised, disappointed, or saddened by it. We take in reports of the carnage caused by natural disasters, tragic accidents,

69

and distant wars with the same dispassion reserved for sports scores or stock figures.

This underscores our profound egotism. We fail to relate to unknown human beings as part of the collective "us." Unfortunately, this failure isn't particularly a modern one. Even Ralph Waldo Emerson writing in1849 hoped human beings would strive to overcome personal ego that excluded others. Emerson wrote: "Culture, the height of culture, highest behavior consist in the identification of the Ego with the universe, so that when a man says I think, I hope, I find—he might properly say, the human race thinks, hopes, finds."[30]

Indeed the basis of many of the world's great religions are founded in consideration, love, and compassion for fellow human beings. As Tolstoy wrote in *The Law of Violence*:

> All the ancient religions recognize that love is the essential condition for a happy existence. The sages of Egypt, the Brahmans, the Stoics, the Buddhists, etc. declared the principal virtues to be kindness, pity, compassion and charity; in one word, love in all its forms. The highest of these doctrines, especially those of Buddha and of Tao- Tse, went as far as recommending love to every human being, and for people to return good for evil."[31]

Tolstoy, who was a devout Christian, went on to say that the

> doctrine of Christ based on the metaphysical principle of love, the supreme law that should guide us in our daily life... should not be considered as entirely new, standing out distinctly from former beliefs. It is only the clearer and more precise expression of the principle that previous religions divined and taught instinctively. Thus it is that instead of love being merely one of the virtues, as it was for (previous religions), Christianity has made it a supreme law, giving man an absolute rule of conduct. The Christian doctrine explains why this law is the highest, and indicates as well the acts that man should or should not commit after having acknowledged the truth of this teaching."[32]

The purpose of this book is not to convince writers to "acknowledge the truth" of a teaching or to embrace a specific religion. Nor is it my purpose to convert writers to a philosophy of nonviolence. Mahatma

---

[30]Whicher, Stephen E. (Ed.). *Selections from Ralph Waldo Emerson, An Organic Anthology*, Houghton Mifflin, Boston, 1960, p. 320.

[31]Tolstoy, Leo. *The Law of Love and the Law of Violence*. Rudolph Field, New York, 1948, pp. 35–36.

[32]Ibid., pp. 35–36.

Gandhi, Dr. Martin Luther King, Jr., and others have written and spoken eloquently in favor of that complex philosophy.

Just how complex is illustrated by the following story related in *The Los Angeles Times Magazine* by Arun Gandhi, founder of the M. K. Gandhi Institute for Non-violence and the grandson of Mahatma Gandhi. He wrote:

> *When I was coming back from school one day, I had this little pencil in my hand about three inches long. I thought to myself that I deserve a better pencil, that this is too small for anybody to use, and so I threw it away. And that evening when I asked him for a new pencil, (my grandfather Mahatma Gandhi) subjected me to a lot of questions. He wanted to know how the pencil became small, and where I threw it away, and when did I throw it away. And then he finally told me to go and look for it, and I thought he was crazy. I asked him "How do you expect me to look for this pencil in the dark?" And he said, "Well, here's a flashlight. Take this and go out and look for it." And I went and looked for it, and spent about two or three hours.*

> *When I finally found it and brought it to him, he said, "Now I want you to sit here and learn two very important lessons. The first is that even in the making of a simple thing like a pencil, we use a lot of the world's natural resources, and when we throw them away, we are throwing away the world's natural resources, and that is violence against nature. The second lesson is that because in an affluent society we can afford to buy all these things in bulk, we over-consume them, we are depriving people elsewhere of these resources and they have to live in poverty and that is violence against humanity."*

> *That was the first time I realized that all these little things that we do every day, that every time we throw away something useful, we are committing an act of violence. So you see the breadth of the philosophy of non-violence. It is almost unlimited.*[33]

My purpose here is to encourage writers to come to terms with what violence (social and personal) means to them. It is essential for us as writers to examine our own beliefs about violence in order to make a decision about how (or if) to use it in our creative work.

How desensitized have we personally become to violence? How does the degree of our own desensitization influence the decisions we make about writing violence into our screenplays? Are we casual about it? Are we careful? Are we callous?

---

[33]Diamond, Nina L. True Believer, Los Angeles Times Magazine, July 29, 2001, p. 23.

Because most people today have been radically desensitized to violence, screenwriters usually think (and are often told by studios) that they must take drastic measures to make an audience care about movie victims and to get emotionally involved in stories. Because most writers don't trust that compelling characterizations and brilliant writing will do the trick, they find themselves depicting graphic acts of violence shocking enough to be sure to get an audience's attention.

That's one of the reasons why the methods of movie murder have grown more diabolical and the results more "eye catching." In the old days when cowboys were shot, they neatly fell from their ponies. Today, we see brains splatter and blood spurt, bodies explode, melt, or burn. More accurate perhaps but much more diverting from the actual point—that the bad guys die in the end. Audiences forget about that as they sit mesmerized, watching the disintegration of a physical body up there on the screen. And because that kind of visual violence is often emotionally charged by characterization and plot development, it takes on even more power on a subliminal as well as on a conscious level.

The power of violence and our increasing insensitivity toward it has tragically manifested in our communities and is negatively affecting young people. Statistics published by the Office of Juvenile Justice and Delinquency Prevention in 1999 say that on the average, juveniles were involved in one quarter of serious violent victimizations annually over the last 25 years.[34]

The statistics go on to tell us that the number of murder offenders in each age group between 14 and 17 increased substantially and proportionately from 1984 through 1993, and although you may not want to make a direct link between media and mayhem, consider that Action–Adventure films are specifically marketed to young males and that one of the givens of this book is that films influence social behavior.

Of course, there are many who take exception to the inclusion of statistics and put down the increase of juvenile crime to factors like overcrowding and family decay. Exactly. Children who feel trapped and have no positive messages about solutions to problems, children who cannot rely on family structure or elders at the top of a community hierarchy, may resort to media for problem-solving tips and, based on violent images inherent there, and the implicit acknowledgment of violence as a legitimate or at least understandable solution in

---

[34]Snyder, H. & Sickmund, M. Juvenile Offenders and Victims: 1999 National Report, p. 63. Washington D.C.: Office of Juvenile Justice and Delinquency Prevention, 1999 and Bureau of Justice Statistics, 1973–1997 National Crime Victimization Survey, data, Washington, DC: BJS, 1998.

times of turbulence, may be compelled to act out their rage and dispossession in violent ways.

Sadly, adults who are the models for children's behavior find that they themselves are deeply involved in the violence of society by becoming victims, perpetrators, protectors, or protestors. Occasionally, adults do take steps to curb violent entertainment. Such action is usually inspired by very real current events that, for all too brief moments, serve to emphasize the volatile nature of our society.

After the 1999 Columbine High School shooting, for example, studios and production companies became more sensitive to teen violence. After the school shooting in Littleton, Colorado, Miramax refused to release *O* (a modern take on Othello that takes place in a high school and shows students shooting teachers). That film was eventually released by Lions Gate Films in September, 2001, right in time for "back to school."

In the wake of the horrific events of September 11, 2001, many entertainment industry executives curtailed production of disaster films. "Some of what we thought was entertaining yesterday, isn't today, and won't be tomorrow," said Amy Pascal, chairwoman of Columbia Pictures. Jon Landau, producer of *Titanic*, summed up the new rules of moviemaking this way: "No bombs on planes, no bombs in buildings."[35]

Finally admitting that films depicting disasters might have some effect on the public psyche, studios and production companies began to revise, shelve, or postpone productions. Whereas terrorists and terrorism were hot before the World Trade Center and Pentagon tragedies, immediately afterward, projects involving terrorism were taboo. The script for Jackie Chan's *Nose Bleed* slated for production by MGM from New Line Cinema was radically rewritten because it originally had Chan battling terrorists atop the Empire State Building. Warner Bros. delayed release of *Collateral Damage*, the Arnold Schwarzenegger film about a man seeking revenge against terrorists who blow up a building, killing his wife and child. James Cameron's production company also stalled *Deadline*, about terrorists taking over a jetliner. Disney delayed the release of *Bad Company*, a Chris Rock–Anthony Hopkins action comedy. Columbia delayed the start date for *Tick Tock*, in which Jennifer Lopez plays a bombing suspect. Imagine Entertainment killed *Flight Plan*, a film about an airport security expert whose daughter is kidnapped on a plane bound for Hong Kong.[36]

---

[35]Eller, Claudia. Hollywood Executives Rethink What is Off Limits, *Los Angeles Times*, Sept.14, 2001.

[36]Goldstein, Robert. A Turn of Events, A Change in Plot, *Los Angeles Times*, Sept. 25, 2001.

Suddenly, bottom-line executives and screenwriters were being sensitive to the emotional sensibilities of an impressionable public. So sensitive, in fact, that studio executives even had the World Trade Center Towers digitally removed from the New York skyline in films like *Zoolander* and *Serendipity*, scheduled for release shortly after September 11, 2001. Personally, I thought that was a creepy thing to do.

Perhaps I'm cynical but I believe that the studios' newfound dedication to ethical filmmaking stems not from their desire to be more socially responsible but rather to make sure they don't alienate audiences shaken by the terrible events of September 11, 2001. Executives are savvy enough to know that glossy Hollywood portrayals of national disasters can make studios and production companies look opportunistic and callous and that makes for bad publicity and poor theater attendance.

"In my experience, whenever anything catastrophic happens, we all vow that the world will never be the same," said Ed Zwick, producer of 20th Century Fox's *The Siege*, a 1998 film about a wave of terrorist attacks in New York. "But one can't live with a heightened sensitivity forever and gradually normalcy returns. That will happen here."

Hollywood critic L. Brent Bozell III, president of the conservative Parents Television Council, believes that "as soon as a little bit of a scab develops on the wound" the industry will even start exploiting the tragedies with TV movies and other projects. Indeed, most of us in Hollywood believe, with screenwriter Jonathan Hensleigh (*The Rock*, *Die Hard With a Vengance*) that "there are TV movies on the drawing board that are re-creations of the events of September 11, 2001."[37]

(Indeed, only a year after 9/11, books and TV movies based on the experience of survivors began trickling into the marketplace, and a rather ghoulish assortment of Twin Towers memorabilia—jewelry, knickknacks, and bric-a-brac—remained on sale.)

Unfortunately, world events also make decisions difficult for writers sincerely trying to be socially responsible in their work. For example, one of my graduate students wrote a thoughtful and artistic short screenplay about terrorism for his MFA thesis film. After the events of September 11, 2001, he came to me and questioned whether or not he should continue with the project. We talked for a long time and I assured him that because his script was thoughtful, intelligent, artistic, and nonexploitative, he should go ahead with the project. It's important to keep events in perspective and to know the difference between

---

[37]Ibid.

exploitation of and capitalization on a tragedy, and an artistic interpretation of tragic events.

It takes wisdom to be able to distinguish between art and exploitation. Unfortunately, Hollywood executives do not have a reputation for
wisdom. They operate on knee-jerk responses. They identify buzz
words and make decisions based on economics. We can't rely on them
to make sound judgments coming out of their sense of altruism, their
understanding of art, or their appreciation of substance. Knowing how
they have operated historically, how can we trust them or go to them for
advice on ethical film making? Clearly, Hollywood executives can't be
considered "elders" in our artistic community. In our current times, the
notion of community elders has become an antique notion and elders
themselves are not easily found.

WHERE HAVE ALL THE ELDERS GONE?

I n the film *Cocoon* a group of old people are conveniently taken off to another planet to continue living full and rewarding lives. Unfortunately, in our own culture, many wish that would really happen. As a society, we've ceased to value age and experience. Many of our institutions—sadly even our own families—do wish that "old people" could be magically removed to a place where they don't demand recognition, attention, or care.

I've already talked about the negative effect on Arctic culture when elders (grandparents, parents, older community members) were superceded by Government. Happily though, in Inuit family structure, elders still command a great deal of respect. The Inuit often live in extended family units where they keep their old ones close to them. The concept of a Community Elder as a wise person who can guide others and provide an example of how to live a moral and ethical life is still strong in Inuit family life.

Our "modern" American communities unfortunately do not function in this way. In current American family structure, elders—people who are perceived as physically or experientially "old"—are often discounted and/or ignored. This attitude is painfully evident in our health care system, in our employment opportunity structure, in our social systems, and certainly, in our media.

It's fascinating to realize that in a world where all of us want to (in the words of *Star Trek*'s Spock) "live long and prosper," we put emphasis on the "prosper" and, though we all yearn to "live long," in our minds we hear that as "live *young*." We've all heard the theory that young people think they'll live forever. I suppose that's true, but I'd also say that young people think they'll "live *young forever*." That thinking has spilled over into most facets of society.

No one wants to admit to aging. Women and men are spending billions slicing and dicing, pushing, lifting, and sucking the effects of age from their bodies. They're injecting themselves with Botulism (Botox) to avoid wrinkling. They're weaving and plugging their heads

76

full of hair. They're reducing, firming, sculpting, working out, and di-
eting like never before. In a land of plenty, people are starving themselves while they gorge on expensive toys and clothes that will bolster
their personal self-images and feed their obsessions for perpetual
youth.

Have they been influenced by Hollywood images? Sure. But they've
also been driven by our basic human need to believe that we'll never
die—that we're immortal and eternal. That pursuit of eternal
youth—as a blissful state where there is an absence of pain and suffering—has become a religion in our culture. That's especially true if you
consider the definition of religion put forth by Paramahansa
Yogananda in his book *The Science of Religion.*

Paramahansa Yogananda said:

> ... *if the motives for the actions of all men are traced farther and farther
> back, the ultimate motive will be found to be the same with all—the re-
> moval of pain and the attainment of Bliss. This end being universal, it
> must be looked upon as the most necessary one. And what is universal
> and most necessary for man is, of course, religion to him. Hence reli-
> gion necessarily consists in the permanent removal of pain and the re-
> alization of Bliss (or God). And the actions which we must adopt for the
> permanent avoidance of pain and the realization of Bliss (or God) ...
> are called religious.*[38]

We'll come back to that definition and deal with the subject of religion in depth later. In the meantime, it's clear that the "religion" of
youth worship is strongly supported by the film industry. Movies are
most often about the young and the beautiful and movies' issues are
most often "youth oriented." Writers are consistently steered away
from writing about old people because "everyone" knows that old age
is a hard sell in Hollywood. The prevailing sentiment has for years
been that older people are not interesting, sexy, or exciting enough to
make people spend money to see them.

Yes, older stars like Clint Eastwood, Paul Newman, Robert Duvall,
Robert Redford, and Michael Douglas are still able to get pictures
made. *Grumpy Old Men* with Matthau and Lemmon did well. But
those are the exceptions. And notice that when older actors do get
leading roles, those geezers are usually paired up with *much* younger
women. (In film as in real life!) Older women, if they even get starring

---

[38]Yogananda, Paramahansa. *The Science of Religion.* Self-Realization Fellowship Press, Los Angeles, 1969, p. 14.

roles, certainly don't get paired up with *much* younger men, and un-like older male actors, older women are asked to keep their shirts on.

Perhaps that's because studios are hiring younger and younger ex-ecutives to make decisions about what gets "green-lighted." They say that younger execs have their baby-fat fingers more on the pulse of tar-get audiences (18–25-year-olds). Maybe they do, but do young and in-experienced execs have the taste, life experience, and knowledge to know what's good? Judging from the kind of stuff that generally gets made, I'm afraid not. But that doesn't usually matter in a world where the bottom-line—revenue—is everything.

That's why older writers (and by older I mean anyone over 40) are having such a hard time selling screenplays. They try to get around this problem in a variety of ways. Here are some more stories from the Horror Vault.

One of my graduate students (at the time only in his early 30s) sold a script and when the sale was reported in *The Hollywood Reporter*, his age was given as 28. When I asked him about that he told me his agent told him he needed to lie about his age to get more work.

Another writer I know with lots of TV credits puts her 22-year-old son's name on the scripts as cowriter and takes him to pitch meet-ings with her. She is no longer surprised when the studio and net-work execs address all their comments to him and ignore her. That phenomenon is not restricted to older women. Older male writers have been doing the same thing, usually partnering up with much younger people or fabricating younger "dummy cowriters" in order to get meetings and assignments.

I was flattered when an 80-year-old writer I met in a bookstore asked me to partner up because he thought I was young enough to get him work. I laughed. To him I was a spring chicken. To Hollywood execs, this hen was past industry prime.

As an aside, women have for years partnered up with male writers because they know that they'll have less trouble getting meetings and sales in a system that tends toward sexism. The problems become even worse when women writers age.

A 55-year-old woman writer I know was once sent into a meeting by her agent for a writing assignment on a love story. The exec she met with was a guy in his late 20s. When she walked in, he looked shocked. Then he carefully explained to her that the story he wanted would have lots of hot sex in it and passion and romantic love and then told her "I can't possibly give you the assignment. What do older women know about things like that?"

Had enough? Well, here's one more and it happened to me. I was be-tween agents and a friend of mine recommended I call a "big agent" he knew. I specialize in comedy and this "big agent" represented a lot of comedy writers. I called him, and things were going well. I started telling him what I'd worked on when he stopped me. "Uh, wait ... how old are you?" I paused. "Hey, that's not a legal question!" I said with as much good nature as I could muster. "So," he grumbled back, "the fact that you're not telling me, and you've had so much experience means you're way way too old to do this. You can send me a sample of your stuff but it doesn't look good." He hung up. Because hope springs eternal, I did send him a script. I never heard from him again!

Age discrimination is so rampant in the industry (especially in situation comedy) that there's currently a class action suit by older Writer's Guild members against producers and studios. The word on the street is that it won't do us much good because agism in Hollywood is the way of life. As far as Hollywood is concerned, the concept of Community Elder—a wiser, moral, and ethical role model—has left the building.

The rest of our society is of like mind. Many of our corporations are asking older workers to take early retirement so that companies can cut down on the larger salaries and perks paid for experience. They reason that younger workers are more eager to prove themselves and so will do more for less money. What they do might not match up to the excellence of experienced workers and younger workers might make horrible mistakes but hey—at least their salaries are bottom-line friendly and they won't get much severance pay if they're fired.

Lots of people who've worked for organizations for a long time have seen how they are often discounted and shoved aside by suddenly popular newcomers who think that success means disregarding the ethics of seniority.

In this ruthless social atmosphere, it's little wonder that our elected representatives—senators, congressmen (Packwood, Condit, etc.), even presidents of the United States (Clinton)—can't be counted on to exemplify impeccable moral and ethical behavior; religious leaders (like Jim Bakker and Jimmy Swaggart) have been exposed as hypocritical, immoral, unethical, and even criminal; police departments (LAPD, NYPD) have been investigated for corruption and violence; sports heroes (Tyson, Strawberry, Rodman) are arrested for violence and drug offenses and/or are implicated in sexual scandals; celebrities (Meg Ryan, Billy Bob Thornton, Jennifer Lopez, Pamela Anderson—far too many to mention) get big headlines for extramarital affairs, divorces, and promiscuous dating, and get congratulated for admitting to drug and alcohol addictions (Robert Downey, Jr., Ben Affleck); and school coaches, teachers, ministers, and priests are being exposed for molesting kids.

Understandably, we deal with the absence of ethical and moral examples in our communities and the violence in our societies by desensitizing ourselves, by isolating ourselves, by denial, or even by grim resignation. Probably, by the time we are adults (and especially if we live in big cities), we don't even realize the extent to which we have become desensitized. We believe that as long as we can live in our own little cocoons of safety, we can avoid the violence that permeates our world.

Exercise One

1. To demonstrate the degree to which you've become desensitized to violence, write a violent scene for a movie you have written or intend to write. Make it as graphic as you can tolerate.
   a. Ask yourself if it was easier for you to write explicit violence than it was to write graphic sex. Be honest.
   b. Write down what message about life your violent scene delivers to an audience.

2. Now, decide if it is easier for you to pinpoint the message that physical violence delivers to an audience than it is for you to determine the message that graphic sex delivers. Keep in mind that reactions to both sex and violence can be quite visceral. They both appeal to our animal nature even though we might call their execution "artistic."

3. Consider how comfortable you would be letting those dearest to you watch your violent scene. If you had children, would you feel comfortable letting them see your violent work?

I know one writer of slasher films who says he won't let his daughters see his work. Does that mean he's comfortable with the idea that other youngsters are seeing it? As far as I'm concerned, this "It's okay for them but not for me" attitude is unethical because it fails to acknowledge that we live in a global community that is affected by our actions and that each one of us bears some responsibility for creating a peaceful world.

4. Answer these questions:
   • Is the degree of violence of the scene you wrote absolutely necessary to the film's thesis and the depiction of character?
   • Will the character's portrayal or the film's thesis be diminished by the reduction or exclusion of the graphic violent nature of the scene?

- What message do you think the scene delivers to the audience about your philosophy?

5. Does the message of your violent scene coincide with your own personal value system?

- What effect, if any, do you think your violent scene will have on the audience?
- Is this the effect you intended it to have? Are you comfortable with the effect you've created?
- Can you take responsibility for having written that violent scene and the effects it might have on the audience with a clear conscience? We'll take a look at what conscience means after the exercises.

Exercise Two

1. Make a list of films and scenes from them in which you feel graphic sex and/or graphic violence are entirely justified. Write down what justifies them. To do this, you will first have to decide if you believe that there's a difference between "good" violence and "bad" violence and what constitutes that difference.

   a. Decide whether any of these scenes could have been shorter or have achieved the same effect in a different way.
   b. Look carefully at the victim(s) of the violence in each of these films. What gender or group did they come from? What statements (overt or subliminal) did the "victim profile" make?

2. Take a graphic scene from one of the films you've listed and rewrite it to be less graphic. You might include camera angles here to add to the impact. The scene might turn out to be even more frightening and gruesome without its graphic aspects.

The shower scene from Hitchcock's *Psycho* (1960) is a great classic example of what can be done with innuendo. Even Janet Leigh said that when she saw herself in that shower scene, she was so traumatized that she's never been able to take a shower since! (Don't worry—she takes baths and is spotless!)

Restructuring a graphically violent scene always takes more thought and skill than it does to write graphically. For example, one of my students recently wrote a script in which a demented and very bad guy tied up the people he was about to kill, and then carefully sliced off both their ears. Then he fed the ears to his pet pit bull who gobbled them down eagerly. The scene was so graphic that it actually made me queasy.

I asked the student if it was really necessary to see the ears actually being sliced. He thought that the slicing was important because it showed the horrid character of the bad guy. Did we have to see both ears sliced? He thought that two ears would indicate that the bad guy was *really* bad. I asked if it was really necessary to actually see the pit bull chomping the ear? He thought that feeding the ear to the dog emphasized that the bad guy was *really really* bad, evil, *and* disgusting.

I conceded that we did have to know that this bad guy was sick and ruthless but I suggested that we really didn't need to see the actual slicing—we could simply hint at that through edits; that the slicing of two ears was overkill; and that just a shot of the guy holding the ear over the drooling dog would suffice. Hey, it was obvious enough that the bad guy was really really bad!

A nice compromise? The student didn't think so. He kept the ear slicing in but agreed that we didn't have to see the dog actually chewing so he only showed the ear being dropped into its waiting mouth! And he said he absolutely needed both ears because he was only going to feed one to the dog and keep the other as a souvenir! (He was after all, really, really sick as well as really, really bad!)

Still too violent as far as I was concerned but I supported the student's right to keep the scene as he wanted it. He'd thought about it, considered it carefully, and was willing to take responsibility for writing it and for its effect on the audience. Remember, it's not about censorship, it's about ethical responsibility!

RECAPTURING INNOCENCE: A RESENSITIZING EXERCISE

Because screenwriters are usually rabid film and television afficionados, they are most likely to suffer the effects of desensitization to violence by media. Lots of us see virtually *every* film that's been released and even some that haven't been. Most of us like to watch a wide variety of film offerings, foreign as well as domestic, even though we may not be partial to the genre of story. We watch films to see what's being written out there just as people in the wine business flock to tastings to sample their competitors' products.

An unpleasant side effect of gorging on film tends to be boredom, increased sarcasm, the development of a hypercritical nature, and, most soul crunching, a laissez faire attitude toward graphic sex and violence. I've heard screenwriters who don't like graphic sex and violence say that they just close their eyes during the bad bits (which have to be pretty bad to trigger that don't-look response). Far fewer writers back away from their film habits by being more selective about the movies they see.

Those of us who do see lots of movies might have become jaded. Can you imagine what it would be like to revisit the experience of seeing your first film? Can you imagine what it would be like to be sensitive to the effects of that film on your heart and mind? If you're game, here's a difficult exercise that might, in some small way, give you back a little of that feeling. If you do it wholeheartedly and are willing to experience change, it will resensitize you to the depictions of violence in media and movies.

It will also demonstrate to you how those who have not yet been desensitized (children), and those who are or who have purposely made themselves more sensitive, might react to movie violence.

Exercise

If you've taken a camping trip to a remote area, gone on a spiritual retreat, taken a vacation to a far-flung corner of the world, or otherwise isolated yourself from society for any length of time, you already have experienced a kind of culture shock when you reentered your usual city life.

The amount of "culture shock" you experience is in direct proportion to the amount of time you've been away and what you've done (or not done) during that time. Coming back from a 2-week vacation isn't quite the same as being on a desert island for 2 months.

Try this *for at least a month.*

1. Even if you live in a big city, and, if your work allows it, isolate yourself from all media sources where violence might be shown. Don't watch TV, go to movies, or listen to radio where news might break in. Do not discuss world affairs or local affairs with your friends or family.

2. During this time, do the following things:

    a. Meditate for at least 10 minutes twice a day.
       Here's how to meditate:

- Sit with spine erect in a chair with feet planted firmly on the floor (or cross-legged if you are limber enough), and the hands with the palms upturned at the junction between the thighs and the abdomen.
- Close the eyes and then turn them slightly upward and concentrate your attention at the point between the eyebrows.

- Every time your mind wanders, bring it back to that point.
- Watch the breath without trying to control it.
- Relax—do not strain.

Interestingly enough, once you begin to meditate, you will discover the true nature of your mind—that it loves jumping around from one subject to the next. Perhaps even the things you've been most avoiding will come up into your thought processes. Simply dismiss them and redirect your attention to the point between the eyebrows.

You might find at first that even one minute feels like an eternity. Never mind. Keep going and don't give up your practice because that's what it is—*practice*—and so it's not always perfect. As you continue, you will notice that your breath slows down and you feel much calmer. You might even begin to feel sensations of joy and of peace. This is an indication that you're on the right track.

You can gradually increase the amount of your practice as you continue.

Because it's important not to recommend doing things that you haven't done yourself, I'll tell you that I've been meditating at least twice a day since 1971 and I've found that it's changed my life for the better and enriched my creative work.

> b. Read only inspiring and uplifting books—no crime stuff, no magazines or newspapers.
> c. Notice your emotions. Notice when you feel angry or frustrated. Make notes of these feelings in a journal. Remember, you are only observing your reactions, not trying to suppress them.

3. At the end of the time you've allotted for this exercise (remember—at least one month!), write a scene in which violence plays a part. Notice how (or if) it is different from the violent scene you wrote in the "Something Shocking" section.

4. Now, re-enter the media world. Pay attention to your reactions during the first day of re-entry as you read the newspaper, listen to news, watch a cop show. If you really want to "test" yourself, watch an action–adventure film that has scenes of violence in it. Record your reactions in your journal. You might be amazed at what you find.

Some of your reactions to the violence you see or hear about might be physical—a clenching of the stomach muscles, a tightening of the jaw, adrenaline rushes, or even anxiety. Perhaps then you might understand the effects even simulated violence or the thought of violence has on our nervous systems.

5. Notice how (or if) your moods change and how (or if) your interactions with others have changed during your time "away."

6. Continue to record your reactions to media and your emotions for at least a week. This will tell you exactly how long it took you to get desensitized again. Was it a day? A week? An afternoon? An hour? The time it takes to desensitize again should indicate to you how powerful the effects of media are and how conditioned you've become to adapting to them. For some, just like an addiction to a drug, watching just one violent film will be enough to get back that "anything goes" and "nothing fazes me" attitude.

7. Now that you're back in the media world, how does your conscience react when you watch violence?

CONSCIENCE

onscience is powerful and influences us in profound ways. I maintain that we always know what our conscience tells us. Some of us may choose to disregard our conscience but when we do that, we often feel unsettled and out of sorts. Only very few people—sociopaths and those with deep social disorders—are able to say they've never experienced any pangs of conscience in regard to any of their actions. When Shakespeare through Hamlet says "Conscience doth make cowards of us all," he's talking about the power of conscience to make us think and that stops us from doing rash and impulsive things that may be unwise.

If your conscience is clear, you're free of nagging guilt—the product of that still, small voice that nudges you into recognizing your mistakes. A clear conscience lets you sleep like a baby, look everyone in the eye, hold your head up, and be proud of yourself and your work. The conscience that tells you if you've behaved correctly or that stops you from behaving badly is the same conscience that tells you whether or not you've done your best and put your all into your writing.

Jacques Maritain, in the Princeton University Council of the Humanities Lectures in 1951, said:

> That which takes place with regard to the moral conscience of man as man is exactly what takes place with regard to the artistic conscience of the artist, as well as with the medical conscience of the physician or with the scientific conscience of the scientist. The artist cannot want to be bad as an artist, his artistic conscience binds him not to sin against his art, for the simple fact that this would be bad in the sphere of artistic values.... To support his family an artist may have to become a farmer, or a customs officer, as did Hawthorne or Henri Rousseau, or even to give up art. He can never accept to be a bad artist and spoil his work.[39]

---

[39]Maritain, Jacques. *The Responsibility of The Artist*. Charles Scribner's Sons, New York, 1960, p. 90.

For Maritain, the same conscience that urges us to be good artists urges us to be good people. He said that artists, because they are human, are very aware of the impact their own moral lives have on their art. He went on to say that conscience demands that artists be sincere with others and, most important, with themselves as they scrutinize their interior lives.

That's because, as Maritian saw it, an artistic discipline *demands* certain virtues that, at times, even imitate the virtue of saints. Those who sincerely want to do exceptional and ethical work subject themselves to

> a kind of asceticism, which may require heroic sacrifices.... (The Artist) must be always on his guard not only against the vulgar attractions of easy execution and success, but against a host of more subtle temptations. He must pass through spiritual nights, purify his ways ceaselessly, voluntarily abandon fertile places for barren regions full of insecurity. From the point of view of the good of the work, he must possess humility and magnanimity, prudence, integrity, fortitude, temperance, simplicity, ingenuousness. All these virtues which the heroes in spiritual life possess purely and simply, and in the line of the supreme good, the artist must have in ... the line of the work."[40]

If we are writers who aspire to this level of artistry, it's understandable how conflicts can arise when we're asked to write something that goes against our personal morals, value systems, or beliefs. Technically, we are capable of writing that kind of stuff but deep inside our conscience will tell us we've made bad art by creating something that, as far as we are concerned, isn't true.

Truth and the search for truth are things all serious writers must concern themselves with. Most secular pundits and all religious leaders would agree that this allegiance to truth is an important human virtue. Quoting from the *Declaration on Religious Freedom* (*Dignitatis Humanae*), Pope John Paul II wrote:

> ... all human beings are bound to search for truth ... and as they come to know it they are bound to adhere to the truth and pay homage to it.... motivated by their dignity, all human beings, inasmuch as they are individuals endowed with reason and free will, and thus invested with personal responsibility, are bound by both their nature and by moral duty to search for the truth, above all religious truth. And once

---

[40]Ibid., p. 99.

they come to know it they are bound to adhere to it and to arrange their entire lives according to the demands of such truth."[41]

To be an ethical writer who wants to create work that is true and satisfying, a writer must write with a conscience that will give warning when personal truth has been transgressed. In order to get in touch with that conscience, writers must sincerely examine their interior lives and determine the personal values that drive their work.

What are personal values? To determine that, each of us must carefully examine what it is in life that has great worth to us and to distinguish between what we take casually and what we truly value.

---

[41]Pope John Paul II, Vittorio Messori (Ed.). *Crossing the Threshold of Hope*, Alfred A. Knopf, New York, 1997, p. 189.

PART III
WHAT REALLY MATTERS

WHAT'S IT WORTH?

opefully, at the end of the resensitization exercise, you will have had a new appreciation of how much calmness and peace contribute to your sense of well-being. Do you value calmness and peace? In today's turbulent world, these things should be of great value. Our society claims to value peace and order and yet most people you talk to crave excitement and diversion. They enjoy drama and like living at a frantic pace.

And, as society is a collection of individuals with different priorities, it delivers a mixed message. We say we value education and want our children to get the best instruction they can, but our teachers receive poor pay while millions upon millions of dollars are heaped on celebrities who do little but appear pretty in public. We say we value honesty, but people who lie (particularly celebrities and politicians) are easily forgiven. We say we value family, but divorce is rampant and infidelities are flaunted and excused. We say we value mutual respect, but bad behavior often goes unpunished. Sometimes, it is even glorified! Society is too quick to forgive some people and then radically punishes others. Nothing's clear-cut.

People are always talking about values—particularly family values—and yet nobody bothers to outline what that means. Are values absolute? Are values culture-specific? That is, are American values different from Armenian values? Are there universally good values? Are values based on religion? These questions require mind-numbing explorations and have no definite conclusions. Each of us has a different idea of what values are.

One of my students writing a *Sopranos* script tells me it's big on family values. He's serious, even though it's all about a mobster. After all, movies about the Mafia (like *The Godfather* and *Wise Guys*) are always touting how great Mafia dons are to their mothers, how loyal they are to their friends, how loving they are to their children. Lots of Mafia members are even regular churchgoers. These guys may kill and maim people but they seem to have great family values!

The Association of American National Advertisers, which promotes the inclusion of family values on television, presented the TV show *Survivor* with an award for promoting these values even though many people think the show is little more than a manual for how to make greed, conspiracy, and aggressive competitiveness work for you.

Although there seems to be no clear, easily agreed on definition of family values, I've come to see (and to remember) that the younger people are, the more defined and passionate they seem to be about their personal beliefs and values. If they believe strongly in absolute honesty, for example, they might refuse to compromise and even cause harm to themselves and others by a brutal and rigid adherence to their belief.

As people grow older, they usually temper their values and even change them either because they've realized that in order to survive today's culture, they need to be more flexible, or because they are profoundly disappointed that the high ideals of their youth couldn't be practically realized. Personal values tend to evolve as individuals evolve and need constant re-evaluation.

So where do values come from and on what are they based? Most people you ask will tell you that values have their base in religion. They say that a belief in God and some religious practice gives one a sense of what values should be operating in life.

Atheists, and even some nonreligious (nonpracticing) believers in God, don't agree. They say that values are humanity-based and the cornerstones of a civilized society stemming from group agreement on social structures. Atheists argue that you can have values and not believe in God.

Perhaps that's possible. Personally, I think it would be pretty difficult to construct a personal value system based solely on social constructs without some notion of a Supreme Being or Eternal Energy. We are all "religious" to some extent, especially according to that definition of religion by Paramahansa Yogananda as bliss-seeking.

Paramahansa Yogananda said that "if we conceive of God as Bliss then and then only may we make religion universally necessary. For no one can deny that he wishes to attain Bliss and, if he wishes to achieve it in the proper way, he is going to be religious through approaching and feeling God, who is described as very close to his heart as Bliss."[42]

If we continue to follow Paramahansa Yogananda's reasoning, we can see how it would become possible to overcome the confines of

---

[42]Yogananda, Paramahansa. *The Science of Religion.* Self-Realization Fellowship Press, Los Angeles, 1969, p. 49.

partisan religious dogma and achieve universal religious unity by virtue of our common human need and condition. Paramahansa Yogananda said:

> ... *if we are once convinced that the attainment of this Bliss-consciousness is our religion, our goal, our ultimate end, then all doubts as to the meaning of multifarious teachings, injunctions and prohibitions of different faiths of the world will disappear. Truth will shine out, the mystery of existence will be solved, and a light will be thrown upon the details of our lives, with their various actions and motives. We shall be able to separate the naked truth from the outward appendages of religious doctrines and to see the worthlessness of the conventions that so often mislead men and create differences between them.*[43]

If writers throw light on the details of the lives of their characters, then it is especially important that before beginning to work, they first throw light on the details of their own lives, their own actions, and their own motives. In this way they can begin to discover what is true for them in order to express it in the work they hope will have universal appeal and impact.

The following exercises will serve to launch this discovery process.

Exercise

1. Put forward your own notion of what religion is.

2. Are you religious? To what degree do you practice and/or use your religion in your everyday life?

Here are some guidelines:

If you believe that your religion guides you in your everyday life and if you like to think that you at least try to adhere to its do's and don't's as far as morality and ethics are concerned, then you might say you are religious.

If you are strict about adherence to your religion, you might consider yourself orthodox.

If you don't give a single thought to your religious practice (except on holidays and maybe not even then), you might still consider that you

---

[43]Ibid., p. 50.

belong to a particular religious denomination even though you say you are absolutely not religious.

If you are one of those people who have decided organized religion is not for them, you may have developed your own religion or even become an atheist.

3. Describe your "daily" practice of your philosophy or/and religion. Be specific.

At this point you may ask why should we even consider religion in a book about screenwriting? Belief in God is a very personal issue. I do believe, however, that we can't address the issue of our own values, until we come to terms with our own relationship to God. It isn't my intention to "convert" anyone to a belief in God. Nor is it the purpose of this book to advance the cause of religious practice. The intention here is to get you, as a human being and writer, to think deeply about the universe and your relationship to it so that you can create characters who seem real and alive and make statements that have real meaning.

In order to do this, you first have to start to examine your own character and that requires a search into how you as an individual find meaning in life. How do you explain the inequities, hardships, and joys of life? What is your idea of how it all works? What is your idea of God?

If you subscribe to a particular religion, these questions can be answered by the precepts of that religion. Things become trickier if you have chosen to reject the religion of your birth (even if that "religion" was atheism) and have not adopted an alternative religion. That's because organized religion provides a defendable philosophy of life and a code of behavior for that life.

If you don't subscribe to any religion, then, it's essential that you come up with your own personal philosophy of life to help you face the uncertainties and crises of human living. Often, people who say they don't belong to an organized religion are hard-pressed to practically apply their personal philosophy because they have never spent much time thinking deeply about it or trying to put it into words.

No matter what philosophy of life you have, as a writer, you need to be able to put it into words. On what is your philosophy based? Is it be based on a belief in a Supreme Being? Is it based on a notion that everything is haphazard and chaotic?

These are serious and big questions, not easily answered, and they can take a whole lifetime to explore—but they are questions that each person must ask himself or herself in order to move forward in life and (though it may not be readily apparent) in creative work. Just as life demands from us a certain approach in order to thrive, so creative work demands that we come up with some definite points of view in the versions of life we've invented.

More Exercises

1. Write down what the idea of God means to you at this moment. I say "at this moment" because your idea of God can constantly change and you should expect it to do so. Don't feel that the determinations you make now will necessarily be written in stone as precepts for the rest of your life.
2. Do you believe that your personal values are based on what you wrote in the previous exercise? If they aren't, then where do they come from? If they came from your parents, then where did your parents' values come from?
3. Now, to define your personal values, make a list of "Value Categories"—the areas with which values are concerned. It's not necessary yet to determine what values are "good" or "bad."

Here's a list of some commonly agreed on Value Categories. You can add or subtract from yours if you like.

1. *Belief*: concerning God or a Supreme Being, Eternal Energy—whatever you feel comfortable calling what some might say is unnameable.
2. *Family*
3. *Love*
4. *Friendship*
5. *Loyalty*
6. *Freedom*
7. *Patriotism*
8. *Honesty*
9. *Laws, Morals and Ethics*
10. *Success*

Now let's examine individual categories by exploring them and answering some probing questions.

1. God
You've already written down what your own relationship is to the idea of God. Now write down how you think that idea works in society.

a. How does an idea of God influence individuals?

- How does it affect communities?
- How does it affect our country?
- Does an idea of God deter crime?

b. If you do not believe in God, then how do you think society would work if everyone else believed like you did?

  - If you do believe in God, how would you imagine that a society run by atheists would work? Because America was founded on the principle that it is "one nation under God," does this mean that America's laws come into question for atheists?

c. Do the Ten Commandments have a place in a totally secular society?

  - Can atheists use the Ten Commandments and if so, how?

d. Make a list of movies and television shows dealing with exploration of faith.

2. Family

Family can mean a great many things to a great many people. There are historical TV families like those in *Father Knows Best, Ozzie and Harriet, The Partridge Family, Eight Is Enough, The Jeffersons, Happy Days, Little House on the Prairie.* In *The Andy Griffith Show*, the family consisted of a single father (Sheriff Andy of Mayberry), his son Opie, Aunt Bea, and Barney, the deputy. Even other residents of the town were looked on as family.

Eventually *Friends* became family, and *Dharma and Greg* became family, and then there's the *Seinfield* family, the *ER* family, the *LA Law* family, the *Frasier* family, *The X Files* family, *The Cosby Show* family ... well, you get the idea. Notice that all of these images of family come from TV and therefore more readily creep into the culture because they become part of our home environments, inserted there repeatedly.

Consider movies that explore family function and dysfunction: *The Great Santini, Terms of Endearment, Ordinary People, The Ice Storm, Welcome to the Doll House*—there are lots of them. Family changes. But maybe the values surrounding the concept of family don't. This value category can include ideas about togetherness (under any circumstance, all for one and one for all?), allegiance, intimacy, and trust.

a. What do *you* mean by family? Create a definition.

b. What does "family values" mean to you in relation to your definition of family?

c. Do you believe all families are dysfunctional? Functional? Both?

d. Do you believe that family bonds should transcend all other obstacles? Is it always "my family right or wrong"?

e. Write down how your definition of family has been influenced by your life experiences.

f. Describe your own family as impartially as you can. Include descriptions of all the groups you've considered as your "family" in your past as well as your present.

g. Consider the value held and expressed by each of your family groups. Did they differ?

h. Did your own values stay consistent or did they change with each family group?

i. Make a list of movies and television shows you find address your particular family issues.

3. *Love*

a. Write down your quick definition of love.

b. Re-examine this quick definition based on your past experiences. You might want to list your experiences in a chronological order noting the effect that each one had on shaping your definition of love.

c. Consider how you would *like* love to be. Write down your idealized definition.

d. Examine what you believe the reality is (perhaps somewhere between the two definitions).

e. Is true love possible? Do you believe that it transcends everything? That it is "holy"? Do you believe love conquers all? Do you believe that it's essential for life?

4. *Friendship*

a. Consider how important friendship is in your life. Do you consider that loyalty to a friend comes before anything?

As an example consider the murder of 7-year-old Sherice Iverson. (As in all true crime cases, I refuse to mention the name of the killer or his associate to avoid giving them attention and/or publicity.) Sherice's killer strangled her in a casino restroom. His best friend saw him holding Sherice down and in spite of that, left the room without saying anything. Even after the killer admitted he had killed Sherice, his friend did not tell authorities.

When he was interviewed on *60 Minutes*, the killer's friend said he did nothing to stop Sherice from being restrained because he didn't know her and didn't think she would die. When he found out the girl was dead, he said he couldn't fathom telling on his best friend. Obviously this is an extreme case, but still it requires you to think about how far you'd go for a friendship.

    b. List the movies that bring up the issues of love and of friendship for you. You will have to make two lists. And whereas it's relatively easy to make lists of movies that are "about love" (all the romantic comedies of course are "about love"), don't cheat by making this too easy. List only those movies that address your personal views of love and/or the conflicts you have about love. Do the same for friendship issues.

5. *Loyalty*

    a. Write down a brief definition of what you think loyalty is.

    b. Is there a loyalty hierarchy—should you be more loyal to some people than to others? To some institutions than to others? To some causes than to others? What is your personal loyalty hierarchy?

    c. What are some instances, if any, in which loyalty can be a negative value?

    d. List the movies that address loyalty issues.

6. *Freedom*

Freedom is a particularly American Value Category. We prize freedom dearly. Our constitution is based on the idea that it is an inalienable right. And yet, the notion of freedom is complicated. Is there such a thing as too much freedom? How does freedom fit in with legal and social restrictions?

    a. Write down your general definition of freedom. Define freedom as it pertains to

- relationships
- work
- social interaction

    b. Do you achieve your idea of freedom in your own life? How?

- psychologically
- physically
- emotionally

    c. Make a list of movies (besides the obvious *Braveheart*) having to do with freedom.

7. *Patriotism*
   a. Write down your definition of patriotism.
   b. Do you believe in "my country right or wrong"?
   c. What are some instances, if any, where patriotism can be a negative value?
   d. Superman always fought for truth, justice, and "the American way." As kids we somehow didn't really question what that was. We just assumed it was goodness. What does "the American way" really mean? Take some time to forge a definition for yourself and to explore that definition.
   e. Does the notion of patriotism fit into your view of the world and of how you see foreign nations? If so, how?
   f. How does (or, does) your notion of patriotism affect how you look at foreign films?
   g. List the movies that address patriotic issues. (These can be foreign or American films.)

8. *Honesty*
   a. How important is honesty to you?
   b. When do you think it's okay to lie? Include here an examination of your beliefs about honesty in business as well as your honesty in relationships. Are these two forms of honesty different?
   c. Make a list of movies with honesty as a central issue.

9. *Law, Morals, Ethics*
   a. How do these tie in with honesty?
   b. Is strict obedience to authority and the law implied in morality and ethics?
   c. Do morality and ethics imply responsibility? If so, in what areas and to whom?
      • Should parents be held responsible for the actions of their children?
   d. If morality is based on religion, can an atheist be moral?
   e. Consider the role *conscience* plays in ethics and morality. We've already looked at what conscience means to artists. Some people believe that conscience is the Voice of God within us. What do you think it is?
   f. List movies in which conscience played a central part.

10. *Success*

  a. Define success.

- Consider how your definition affects your judgment of others and of yourself.
- How does success affect behavior? (Consider the behavior of successful people and also the behavior of others toward them.)

  b. Define failure.

- What does failure mean to you?
- What are its characteristics?
- How does it affect behavior? (Also consider the behavior of failures and the behavior of others toward them.)

  c. How much does your idea of success have to do with money? You've already looked at the issue of money in the chapter on motivation. Now take a closer look by determining how much money really means to you.

- How does money affect your decision-making process?
- What will you sacrifice in order to have the quality of life you want?
- What part do you think money plays in our society?
- How do you conform with and/or differ from how our society looks at money?
- Try to forge a *realistic* statement of what money means to you.
- Has money ever caused friction or conflict for you within any of the value categories?

  d. Make a list of movies that deal with success issues.

The exercises you've just done may tell you what some of your beliefs and "values" are. But just knowing that doesn't guarantee that you will develop a consistent code of behavior based on them. That needs a conscious and informed commitment on your part. And in order to make that kind of commitment , it's naturally important to be clear about the nature of your values. That's because, consciously and even unconsciously, you'll end up communicating your values in everything you write.

THE GOOD, THE BAD, THE BLURRY

ow that you've explored what you believe and value, you also need to decide if those beliefs and values enhance and improve your life and the lives of those around you. By carefully scrutinizing our personal values we can come to realize how we are self-destructive or life-affirming. And a closer investigation of what effects our values have will help us determine and define how particular characters behave and react in the screenplays we are writing.

Negative values will have a negative effect on our lives and there's no denying that! How do you know if your values are negative? Here's how you can decide.

෯ Consult your conscience.
෯ Analyze the consequences the application of your values have on your life and on the life of others around you.
෯ Compare your value system to the value system of those you respect.

If you don't like your values—if you've found they don't work for you—then change them. It *is* possible to change values by consistent, continual vigilance. It's not easy but it can be done. Bad values can be addicting, and like other addictions, become habits we need to root out by will power and effort. There are 12-step programs out there for nearly everything. Maybe there's even one out there for the "bad" value you hold that you want to change. Liars Anonymous? Betrayers Anonymous? Gossips Anonymous? Maybe not. But you might be able to fit into a program that comes close. In any case, you can privately adopt the precepts of a 12-step program and make them work in your own life or try psychotherapy to help get the results you're looking for.

What values work? I've purposely stayed away from pronouncing what values are "right." That's because I don't like being told what to

think and assume that you don't either. Alan Wolfe, director of the Boisi
Center for Religion and American Public Life at Boston College wrote
that "we live in an age of moral freedom, in which individuals are expected to determine for themselves what it means to lead a good and virtuous life. We decide what is right and wrong, not by bending our wills to
authority, but by considering who we are, what others require and what
consequences follow from acting one way rather than another."[44]

Wolfe assembled a research team and talked with people from all
walks of life about what it meant to lead a good and virtuous life. Interestingly, he discovered that:

> the desire of so many Americans to have a greater say in the moral
> choices they make is anything but a bitter renunciation of religion. It
> is more likely to take the form of a prayer that someone in a position of
> religious authority will take them seriously as individuals with minds
> and desires of their own. Far from being secular humanists, Ameri-
> cans want faith and freedom simultaneously.... It suggests that in
> America, religious institutions will not break under the weight of
> moral freedom but bend, as many of them have bent already, to ac-
> commodate themselves to the freedom of moral choice to which Amer-
> icans have increasingly grown accustomed.... No matter how strong
> their religious and moral beliefs, nearly all people will encounter situ-
> ations in which they will feel a need to participate in interpreting, ap-
> plying and sometimes redefining the rules meant to guide them."[45]

Even religious leaders like the Dalai Lama recognize this need for
interpretation and redefinition. The Dalai Lama urges us to make decisions with reason and compassion. And yet he counsels if "we are to
retain our peace of mind and thereby our happiness, it follows that
alongside a more rational and disinterested approach to our negative
thoughts and emotions, we must cultivate a strong habit of restraint in
response to them. Negative thoughts and emotions are what cause us
to act unethically." In recognizing our need for freedom the Dalai Lama
went on to say that

> some people feel that although it may be right to curb those feelings of
> intense hatred which can cause us to be violent and even to kill, we
> are in danger of losing our independence when we restrain our emo-
> tions and discipline our mind. Actually, the opposite is true.... When
> we indulge our negative thoughts and feelings, inevitably we become

[44]Wolfe, Alan. The Final Freedom. *The New York Times Magazine*, New York,
March 18, 2001, pp. 50–51.
[45]Ibid., pp. 50–51.

accustomed to them. As a result, gradually, we become more prone to them and more controlled by them.[46]

As we redefine, interpret, and apply, by deep introspection and reflection on our experiences, most of us can and will adopt positive values that will improve our lives, and we will strive to do so because, again on an intuitive level, most of us know what positive values are and recognize that they are profoundly necessary to a healthy personal life and a healthy society. This is an ecumenical process that includes practitioners of all religions as well as secular humanists and atheists, because it involves conscience.

As the American Yogi Sri Daya Mata said in her book *Finding the Joy Within You*, "The survival of civilization depends on observance of standards of right behavior. I am not talking about man made codes that change with changing times, but about timeless universal principles of conduct that promote healthy, happy, peaceful individuals and societies—allowing for diversity in a harmonious unity."[47]

Unfortunately, modern "political correctness" usually deters writers from making public statements in favor of "positive" values, just as it makes our schools shy away from teaching specific values to kids and discourages punishment for inappropriate behavior or negative values held by adults. Education theories and cultural practices since the late 1960s have promulgated the idea that all of us (and especially young people) must be granted the "space" and ultimate freedom to make our own decisions about everything—even ethics and morality.

Educator Dr. Harold Taylor in his monograph *Art and the Intellect* talked about the difficulties that arise because of that. He said that

There is a psychological as well as social sophistication in young people who have been brought up with enlightenment, who understand the arts, who have received the psychological understanding of parents, teachers and everyone else. This combines to produce a political conservatism and an acceptance of social reality and the kindness of other people just as they stand, without a wish to change society in any way but only to criticize it and to criticize such things as "them," mass culture, and mass education.

[46]Lama, Dalai. *Ethics For The New Millennium*, Riverhead Books, New York, 1999, pp. 97–89.

[47]Mata, Sri Daya. *Finding The Joy Within You*. Self-Realization Fellowship Press, Los Angeles, 1990, p. 30.

The dilemma for this sophisticated student is that often he is too intelligent to accept the values of his society, yet he needs desperately to belong to a community to which he can be loyal. He therefore rebels against the authority, and at the same time he is not quite sure he wants to go on being a human hatpin. He has been taught to be critical of established values, but he has become tired of being himself all the time.

We have developed in this progressive environment, a student who is unable to rebel productively because there is nothing to rebel against. He is asked all the time to be himself. A great deal of the time the self he is asked to be is so unformed that he doesn't know what it is or how to be it.... On one hand he asks for guidance and authority and intellectual discipline; on the other, he won't accept anything anyone tells him on the grounds that (what is told) is just the way of authority and that (it's) just discipline. He believes in the right of personal decision, free choice, and free speech but only for other students, not for teachers or parents."[48]

That's true of all of us. Not only students, but parents, teachers, and especially writers suffer from these kinds of difficulties. How can things be changed?

Sri Daya Mata said:

The present trend of permissiveness needs to be turned around and one way is to provide proper training in the formative years. Children should be taught right moral attitudes and right behavior—not only through words, but example as well. Lack of such guidance is a major factor in the tragic breakdown of moral standards and behavior in this country, which has done more than anything else to destroy the family unit.

And what has that brought forth? Emotionally crippled children. And emotionally crippled children generally become emotionally crippled adults, who have developed a feeling of rejection, which leads to bitterness toward society as a whole. They feel that the world has not given them their just due. If not corrected, this breakdown of morale can result in a deterioration of moral responsibility such as that which led to the decline and fall of past civilizations.[49]

---

[48]Taylor, Harold. *Art and the Intellect, The Museum of Modern Art,* Doubleday (distr.), New York, 1960, pp. 32–33.

[49]Mata, Sri Daya. *Finding the Joy Within You.* Self-Realization Fellowship Press, Los Angeles, 1990, p. 176.

As writers (in every form of writing but particularly in popular art), we are especially positioned to act as community elders, and because of our unique platform we owe it to ourselves and to the world to do that by having the courage to take moral responsibility, by holding up certain moral standards, by creating models of right behavior (we'll talk about that in depth in our chapter on Angelic Acts and Vile Deeds), and by consistently working toward a positive and life-affirming value system.

Are these aims too lofty for writers? Certainly not! Walt Whitman even went so far as to call the best and highest form of writing a form of religious expression. In a footnote to his essay *Democratic Vistas* he wrote

> *The culmination and fruit of literary artistic expression, and its final fields of pleasure for the human soul, are in metaphysics, including the mysteries of the spiritual world, the soul itself, and the question of the immortal continuation of our identity. Here, at least, of whatever race or era, we stand on common ground. These authors who work well in this field—though their reward, instead of a handsome percentage or royalty, may be but simply the laurel crown of the victors in the great Olympic games—will be dearest to humanity, and their works, however aesthetically defective, will be treasured forever. The altitude of literature and poetry has always been religion—and always will be."[50]*

This is perhaps the ultimate argument for the importance and impact of content in a writer's work. Certainly it is a clear plea for the adoption of the highest moral and ethical values by those who write.

Exercise

We've got to learn positive value systems somewhere. Hopefully, each of us has had the opportunity at some time or other to come in contact with people who have modeled these positive values to us.

1. Make a list of those people (currently living) who you know model positive values.
2. Interview as many of them as possible. Don't be concerned if some of the people on your list are celebrities or other strangers. I've learned from my years of experience as a reporter that you can talk to just about anyone by simply asking. If you tell people

---

[50]Whitman, Walt. (Ed. James E. Miller Jr.). *Complete Poetry and Selected Prose.* Houghton Mifflin Company, Boston, 1959, pp. 495.

that you're a writer interested in exploring values and ethics, you'll be surprised at how many of them will be willing to talk with you.

3. Ask your interview subjects:
   - How they came about their values
   - How they make crucial decisions
   - What their faith process entails

4. Ask one of these people to mentor you either in person (occasionally meeting) or if that is not possible, through correspondence.

5. Refer to your interview notes and make contact with your mentor when you are faced with moral, ethical, or conflicting value dilemmas. These will occur even though you've already worked hard at defining your personal values.

The next chapter gives you some models for making difficult ethical decisions without consulting eight balls or tossing coins. And whereas using these models might help you in your personal and writing life, they'll be of definite help when you map out the choices each of the characters in your screenplay will have to make to propel your story forward.

NOTHING LEFT TO CHANCE

nce you've determined what values you hold, you must make a commitment to them. That commitment can only be demonstrated by the practical application of your value system in your life and your work. Ultimately, that practical application depends on choice and making the "right" decisions—a process that can be very difficult, particularly when two conflicting values come into play.

For example, the Unibomber was turned in to police by his brother. In all probability, the Unibomber's brother was horribly conflicted by his duty to humanity and his duty to and love for family. How did he make his decision? He ultimately decided to sacrifice the smaller good for the larger good and became a real hero. I'm sure his conscience played a huge role in his decision.

In dramatic situations, however, it's not a good idea to justify the decision a character makes based simply on conscience. We've got to be able to show that character's decision-making process in ways that will be believable to the audience. Saying "God told me to do it" or "the devil made me do it" just doesn't cut it. We've got to come up with original ways of demonstrating decision making if we're going to involve modern audiences. And of course that process begins with first realizing that decision making comes out of internal conflict

And by "internal conflict" I mean the realization that several options for action exist simultaneously and that, just as we individuals have choices and use our values and philosophies to make decisions, every character we create has a choice.

In fact, usually it's the decision-making process based on internal conflict that creates the action in our screenplays. But the outcome of internal conflict most often gets portrayed as violence or other antagonistic forms of external expression because it's easier to show than the often more interesting internal processes. Movies come to a grinding halt when characters talk about what they are going through instead of demonstrating it visually. "Talky" ethical decisions and "talky" self-reflection on choices often become preachy.

A graduate student I was working with wanted to "show" his main character making a difficult moral decision. He thought he could avoid being preachy simply by having other characters comment on that decision. I made the student's job more difficult by asking him to find another way of demonstrating reactions to the main character's decision. Just because his main character wasn't preaching by describing and defending his decision, didn't mean that the script wasn't preachy. It was preachy because of the discussions his other characters were having about the decision.

The truth is that unless you can find some visual way of showing characters "deciding," if you use talk, you are preaching. If you think hard and refuse to take the "easy" way out, you can come up with all kinds of clever and visual ways to show the decision-making process and its effects. And one of them is NOT having the main character walk around his room beating his forehead and muttering "What to do? What to do???" We'll explore ways of making that process visual in a later chapter.

But first, before we show a character dealing with an ethical dilemma, we've got to be able to know how to define an internal ethical conflict and come up with ways in which it can be practically approached in the writing process.

To do that, we'll consider five traditional models of ethical decision-making processes and show how these can become valuable tools to use not only in our personal lives but also in defining and practically resolving our characters' ethical conflicts. These models (called *ethical guidelines* by Christians, Fackler, Rotzoll, and McKee in their book *Media Ethics: Cases and Moral Reasoning*) are taken from history, philosophy, and communication theory.[51] They've never before been used in relation to screenwriting and yet work very well when that's done.

Keep in mind though, that we've got to take some creative leeway with the models. Donald Wright explained in Salwen and Stacks' *An Integrated Approach to Communication and Theory Research*, that "four criteria form the basis of any system of ethics and these are shared values, wisdom, justice and freedom. Ideally ethics and moral values outline the ideals and standards people should live by."

---

[51]Christians, Clifford G., Fackler, Mark, Rotzoll, Kim B. *Media Ethics: Cases and Moral Reasoning*. Fifth Edition, Longman, New York, 1997, pp. 11–12.

However, Wright conceded that "no set of principles exists that will solve all ethical dilemmas."[52]

Using these ethical models as tools for decision making may therefore involve combining the best parts of the models; taking only that which works in a particular situation and applying the principles of the models selectively. And because characters change throughout the course of a screenplay, several models might be applied in the course of the character arc. We'll see that at work when we examine practical application of the models, writing process applications, and, finally, films that exemplify each model.

ETHICAL DECISION-MAKING MODELS

I'm going to streamline these models for you and make them as accessible as I can.

Aristotle's Golden Mean

Aristotle said that virtue lies between two extremes and advocated moderation in everything, including morality. He said that someone of moral virtue will act with careful control. In yoga philosophy, this notion of balance is often referred to as "walking the razor's edge." Extremes are to be avoided. Actions are to be carefully considered. This means that generosity comes between stinginess and wastefulness; modesty between shamelessness and bashfulness; courage between cowardice and recklessness.

Practical Life Applications

◞ Keeping secrets.

- You discover your best friend's husband is having an affair. Rather than tell your friend, you might talk to the husband and urge him to change his ways.
- You discover your employer is defrauding the public. You desperately need to keep your job. You might go to your employer, tell him what you know, and urge him to stop his evil ways. More realistically, you need to quit, get another job and then, in the name of honesty and to keep other cus-

---

[52]Salwen, Michael and Stacks, Don. (Eds). *An Integrated Approach to Communication Theory and Research*, Lawrence Erlbaum Associates, Mahwah, New Jersey, 1996, pp. 525–526.

tomers from being defrauded, you could report the fraud to the authorities.

๛ Disaster strikes.

- (From a true story) You are in the lead of an around-the-world sailing race in open ocean. You have been desperately trying to win this race for many years. This is your last shot. On the radio, you hear that one of the other boats in the race is in trouble and going down in a storm heading your way. You are closest to the boat (and in these waters that means 200 miles away!) and the only possible rescue available in the current weather conditions. You have a limited fuel supply and need to keep moving if you want to outrun the storm and win the race. What do you do? In real life, the woman leading that race went off course and made the rescue because, after analyzing her probabilities, she felt she would be successful.

The Writing Process

Writers using the Golden Mean will:

๛ Make characters more realistic by tempering extreme behaviors. Villains will be more humane and heroes will be slightly flawed.

๛ Ensure that characters' actions are properly motivated. Characters don't just up and do something for the hell of it. They usually think things out and/or act on proclivities of their personalities. Their behavior will be consistent with their biographies and profiles and tempered by their back stories.

Movie Example

Jurassic Park (1993)
Written by: Michael Crichton and David Koepp
Directed by: Steven Spielberg

Jurassic Park III (2001)
Written by: Peter Buchman, Alexander Payne, and Jim Taylor
Directed by: Joe Johnston

These movies grapple with some pretty heavy bioethical questions. In *Jurassic Park* they're voiced sometimes too clearly by Dr. Ian Malcolm (played by Jeff Goldblum), who acts as a kind of Greek chorus warning everyone about the dangers of fooling with Mother Nature. He may be the voice of reason, but the ethical decisions made by Dr. Alan Grant (played by Sam Neill) are most illustrative of the Golden Mean model.

Dr. Grant is in the middle of an exciting paleolithic dig with his kid-hungry botanist girlfriend, Dr. Ellie Sattler (Laura Dern), when he's lured into taking a consulting trip to Jurassic Park. He agrees to go only because he's promised an additional 3 years of funding for his work—something he's ultrapassionate about. Cautious and dino-savvy, he's very careful in everything he does, and although he's enthralled by the Park's magnificent beasts and respects their splendor, he's able to use his knowledge of their behavior to save himself and the kids he resists liking.

Grant always thinks carefully before he acts and plans his moves. He takes stock of situations and makes sure he's made the best decision possible. And he winds up an unlikely hero. In *Jurassic Park III* he's at it again. In this sequel he's tricked into going back to the park and because of his superior knowledge and caution, winds up again saving everyone.

The Golden Mean model is one of the most difficult to apply to movies because it's the most undramatic. All about moderation, caution, and reflection, it's not easily translated into film unless, the character using it must walk a tightrope through pandemonium. In the *Jurassic Park* movies Grant does just that, using the scientific knowledge that by now has become second nature to him to avoid death at the claws and jaws of some very dangerous and burly beasts!

Sometimes, the Golden Mean Model, because of its reliance on caution and personal conscience and expertise, can blend with the Bok Model, which takes much more movie time to implement. (You can see this happening most obviously in the Bok Model Movie Example #2, *Thirteen Days*.)

The Bok Model

Sisela Bok was a philosopher who suggested that when making an ethical decision you should:

- consult your conscience
- consult an expert

◦◦◦ have public discussions with the parties involved to determine the effects of potential decisions

The Bok Model makes use of conscience and the advice of trusted elders, that is, people with experience in the field. It also encourages logical thinking in which each decision is analyzed in terms of its consequences and effects.

Practical Life Application

◦◦◦ Your mate tells you he wants you to do something that will negatively affect his parents. They know about it and secrecy is not an issue. Your conscience tells you this is a bad thing to do. You can consult a therapist. You can also talk it over with your mate's parents to see if they really will be negatively affected. You can then make your decision based on what the process has helped you decide.

◦◦◦ Your friend asks you to invest in her company in a manner that you feel is rather shady. Before you agree, you decide how important the friendship is to you, and whether you can afford to lose the money. Then you consult an investment counselor and a lawyer.

◦◦◦ You discover that the health of your community is being endangered by an industrial plant in your area, but you know that this plant employs most of the people in the community and if it shuts down many families will face financial disaster. To decide what to do, you might consult lawyers, health professionals, and go to the plant executives and to the members of the community to discuss the situation.

The Writing Process

You aren't sure about how to solve an ethical or moral dilemma in your script. You know you can create multiple versions based on different approaches. For each possibility you can:

◦◦◦ Research the issue and consider how these dilemmas have been solved by others. Go to other movies, historical documents, biographies, and public records.

   Think about how writing each solution will make you feel about yourself and your work. Decide which feeling you'd most like to live with.

   Consider what public impact each version of your script will have.

   Enroll in a writing course or hire a script consultant and get the opinion of the instructor.

   Discuss your work in a writers' group and poll members' reactions.

Movie Example #1

Erin Brockovich (2000)
Written by: Susannah Grant
Directed by: Steven Soderbergh

...based on a true story about the real Erin Brockovich, a feisty single mother (played by Julia Roberts) who dresses like a hooker and thinks like a lawyer. Unemployed, uneducated, and desperate, Erin tries to get a job to make ends meet by convincing her own lawyer (played by Albert Finney) to hire her. Messing up in the job, she's almost fired but is able to convince her boss that a suspicious real estate deal she's investigating is worth checking out. When she does dig deeper, she discovers that the issue is bigger than she thought—Pacific Gas and Electric is dumping toxic waste and poisoning families who live on that land.

At first, motivated primarily by her need for money, Erin works to unearth vital information in the case. But the more she investigates, questions people, and consults experts, the more she is moved by the terrible unethical practices of the Big Company. She ultimately convinces her reluctant boss and the suffering townspeople to fight the Big Company in court. In true Hollywood tradition (and happily in real life, too), they win the case.

Erin's decisions are specific to the Bok model because she makes them in the course of a stringent investigation where she asks questions, researches documents, consults experts, interviews "victims," and finally follows the dictates of her conscience to help them.

Movie Example #2

Thirteen Days (2000)
Written by: Philip D. Zelikow
Directed by: Roger Donaldson

Based on a true and tense historical event, *Thirteen Days* is an account of the Cuban Missile Crisis of 1962. For nearly 2 weeks, President J. F. Kennedy (played by Bruce Greenwood) and a close staff of advisers (the closest of these, a fictitious Kenny O'Donnell played by Kevin Costner) play "who'll blink first" with the Soviets, and bring the world to the brink of a nuclear war.

Throughout this dangerous and tense situation, the President consults with advisers, carefully weighs all the options open to him and the earth-shattering results each of these options might have, and ultimately acts decisively according to the dictates of his own conscience.

The decisions Kennedy makes in the film are classic examples of the Bok model, which encourages the exercise of deliberation, caution, and careful study in conjunction with inner inclinations. And it is important to notice that in spite of his caution, the President never once looks cowardly or indecisive. In fact, his awareness of the gravity of the situation and the self-restraint he shows in the face of the hawks around him who urge an immediate military solution, make Kennedy the central heroic figure of the film.

Kant: The Categorical Imperative

Philosopher Emanuel Kant said:

- We should act on the premise that the choices we make for ourselves could become universal laws.

- There is an absolute right and an absolute wrong. Extenuating circumstances don't matter. We must do what is right no matter what.

- Conscience tells us what is right and so we must absolutely obey our conscience. Duty figures greatly in this model. Kant insists that each one of us is honor bound to do the right thing whether we like it or not.

Practical Applications

The Unibomber's brother felt it was his duty to turn in his sibling. Even though he knew that his brother was mentally ill and even though he loved him and didn't want to hurt him or their mother, he knew that he was honor bound to stop the Unibomber from continuing his acts of violence. His conscience and sense of what is right would not let him keep his brother's whereabouts a secret.

(A true story) The mother of a boy who was molested confronts her son's attacker in court and shoots him. She operated under the categorical imperative, but so did the state by putting her in prison for murder.

The Writing Process

Writers who use the Categorical Imperative will staunchly defend their points of view, often refusing to make changes they consider might destroy the integrity of their screenplays. These kinds of decisions are painful and problematic and we discuss them further when we talk about Utilitarianism.

- Writers who use this model will take very definite approaches with story lines and characters. They will not be wishy-washy or inconclusive.
- Writers will do what is necessary to make their screenplays work. And some writers might use the Categorical Imperative rationale to do whatever is necessary to make their screenplays sell if they believe that it's the right thing to do. (Isn't that a grim thought?)

Movie Example #1

The Fugitive (1993)
Written by: Jeb Stuart and David N. Twohy
Directed by: Andrew Davis

Dr. Richard Kimble (played by Harrison Ford) is convicted of killing his wife. He claims he is innocent and escapes on his way to prison. He's relentlessly pursued by Deputy Marshall Sam Gerard (played by Tommy Lee Jones). Finally cornered after a breathtaking series of chase sequences, Kimble decides to appeal to Gerard's sense of fair play. Exhausted and wringing wet, his back to a waterfall, Kimble shouts to Gerard, "I didn't kill my wife."

Gerard, also exhausted, shouts back, "I don't care." The pursuit continues. Following Kant's categorical imperative, the Marshall believes that he must under any circumstances do his duty by capturing a convicted killer.

Movie Example #2

The Contender (2000)
Written and Directed by: Rod Lurie

Senator Laine Hanson (played by Joan Allen) is going to be nominated as the first woman candidate for vice president. Searching to discredit her, opponents come up with information that while she was in college, she participated in questionable sexual antics. Her supporters urge her to defend herself against the charges but she refuses even though the negative accusations make a mess of her candidacy and her life. Because she believes that investigations into candidates' private pasts are improper and that nominating committees do not have the right to ask certain questions, she is willing to go down to defeat to stand by her principles.

Her choice to do so is an example of the categorical imperative model: She does what her conscience tells her is right, she ignores extenuating circumstances, and she does her duty to herself and to her country, assuming that any action she takes will set political precedents that could become election standards.

John Stuart Mill: Utilitarianism

The consequences of actions are important in deciding whether they are ethical.
For example, it may be necessary to hurt one person to save the group. The ethical choice will benefit the most number of people.

Practical Applications

The horrible tragedy of September 11, 2001. Passengers on hijacked Flight 93 made the decision to rush the hijackers. In making that heroic decision, they sacrificed themselves but potentially saved many more people from destruction.

The Utilitarianism model is also used by the U.S. military. President Bush has empowered the U.S. military to shoot down any hijacked commercial airliner that threatens to repeat the terrorist acts of September 11, 2001. It's a horrible thing to do and yet it might be necessary to sacrifice a plane full of people in order to save thousands of others on the ground.

Writing Process

Utilitarianism meshes nicely with the Categorical Imperative. We may have to sacrifice characters or scenes we like in order to present our thesis more clearly, in order to pay off our main idea, or in order to make our screenplay more accessible or unambiguous. I've had students tell me that they will not take an extraneous or bloated scene out of a screenplay because they "are in love with it." Too bad. Sometimes it's necessary to sacrifice your small loves so you can hold on to your greatest one.

Eventually, with experience, these kinds of writing sacrifices become insignificant. The more difficult writing sacrifices have to do with those that are dictated by agents or studios who want to make your screenplay more commercial. Often writers are asked to do terrible and mindless things to their screenplays to make them sell. But be careful not to anticipate these kinds of demands by turning your screenplay into mindless commercial drivel all by yourself. Wait until they make you do it! Ultimately, if you're paid for your scripts, you must make the changes demanded or have your script rewritten by a stranger. And chances are that it will be rewritten no matter how much you try to please those who pay you. Take heart in knowing that you can have your name removed from the credits if you don't agree with changes made to your work.

Writers who refuse to compromise their work can take comfort in knowing that at least they didn't participate in their own self- immolation. But in the end they might still suffer. Because producers and directors won't work with writers who give them grief, writers who refuse to make changes might be committing professional suicide. That's why it's important for writers to take a stand *only* when their personal ethics and values are being compromised and, even then, to consider if the sacrifice they're about to make is worth burning bridges over. Writers shouldn't fight over small stuff when they need to save their strength for big battles.

Movie Example

The Insider (1999)
Written by: Eric Roth and Michael Mann
Directed by: Michael Mann

Research Biologist Jeffrey Wigand (played by Russell Crowe) sacrifices himself for the good of the public's health by talking to *60 Minutes* about the unethical and ruthless practices of the big tobacco company that once employed him. He and his family get death threats, his life is disrupted, he is dunned and followed, his wife leaves him, he suffers financial ruin—all this even though *60 Minutes* refuses to air his interview because it gives into bullying by the network. Yet in spite of it all, Wigand continues to maintain that his self-sacrifice is worth it. Later, we'll examine in depth the choices Jeffrey Wigand makes using the Utilitarian model.

The Utilitarian Model is often seen in combination with the "Love" Model where sacrifices (one for many or one for one) are made out of personal or impersonal love for other people.

The "LOVE" Model

Textbooks call this the "Judeo–Christian" model and, at the same time, point out that the notion of "brotherly" or unconditional love is not exclusive to Western doctrines. As we've already discussed, every religious philosophy involves love. In this model, love is the motivating force for ethical decision making. And it is a love that includes loyalty, unconditional acceptance and friendship, commitment, constancy, and fairness.

Practical Life Application

Practically, this model is most applied in romantic love and in friendship and family matters.

Ask yourself:

- How far would you go (what sacrifices would you be prepared to make) to defend a family member?
- What danger would you put yourself in to protect a friend?
- What would it take for you to be disloyal to a friend or loved one?

Writing Process

Writers who operate under this model will honor their contracts and their partnerships. They will be honest in all their transactions. They will respect promises and friendships. A practical example? Here's a personal Horror Vault story that will illustrate the breakdown of the Love Model.

I wrote a treatment for a miniseries and partnered up with a producer. We had a letter of agreement that said we would share in all profits and participation in the project and that I would be the writer of record. At a high-level pitch meeting (with the vice president of production at a major studio), the executive in charge asked my partner (remember, I was in the room) why she needed me now that she had the treatment. In fact, the executive encouraged my partner to drop me, get another writer, and produce the project herself. I sat open-mouthed and stunned as I heard the executive tell my partner that our letter of agreement could be dissolved.

My partner was desperate to sell the project so she tried to do what the executive suggested. The Love Model was not used here! Had my partner and I made an agreement to sell the project no matter who got dumped, our partnership might have "worked"—at least the project might have gotten made and I would have been prepared for her desperate actions and given a free hand to do the same thing to her if I had to.

Sadly, loyalty to writing partners in the face of studio and production pressures becomes difficult. The desperate desire to get a project made or to get a script sold has destroyed many writing teams and in the process ruined marriages, broken up lifelong friendships, and created horrible silences even between brothers and sisters. If you are going to get involved with a writing partner, the best thing to do is to be entirely honest about the depth and breadth of your "love." It's not such a bad thing to admit to each other that if it becomes necessary to do so, you'll sell out each other. That way, with everything out in the open and agreed on, there will be no surprises or hurt feelings. After all, complete honesty may be one of the more important aspects of "love."

Writers who use the Love Model will refrain from writing stereotypes, will be fair to minorities, and will refuse to include gratuitous scenes of violence and degradation because of their love and respect for others.

Movie Example #1

Titanic (1997)
Written and Directed by: James Cameron

Because lifeboats are in short supply, men stay on board a sinking cruise ship while women and children are loaded onto the boats. (Of course, some "cowardly" men take the lifeboats, but in general men sacrifice themselves so that the women and children might survive.) (Utilitarianism comes into play here.) Using Utilitarianism and the Love Model in combination, Jack (played by Leonardo Di Caprio) sacrifices his own life by letting Rose (played by Kate Winslet) float on top of the water on the only bit of debris available while he stays in the water and dies from exposure.

Movie Example #2

Thelma and Louise (1991)
Written by: Callie Khouri
Directed by: Ridley Scott

Two women who are best friends go on a thrill ride across the country. Egging each other on into more and more daring exploits, they end up committing crimes and finally kill themselves in the name of camaraderie and friendship. The women make ethical choices based on friendship and love for one another in that friendship.

Love itself is, of course, a big motivator for ethical choices in the Love Model, but if you think carefully about it, you'll notice that love—albeit misplaced—is also a very strong motivator in making unethical decisions. People can be moved to commit foul deeds in the name of love and self-love; for the love of money, power, fame. If you take that into account, then in the broadest sense, perhaps love is the underlying motivation of all of our actions and the most profound model of all.

Following that reasoning, keep in mind that these models categorize and define ethical decision making. They have little to do with whether the decisions made were "right" or "wrong," ethical or unethical. That distinction is more involved with our concept of what is Good and what is Evil. We'll examine those concepts in a later chapter. Right now, as we've explored methods for making ethical decisions, it's time to think about creating characters that will have to make these choices.

WHO CHOOSES?

Every writer knows that you can't have a full and satisfying story without characters. Plot twists and sequences, no matter how interesting or intricate, don't give substance and depth to a story. In fact, stories that are plot-heavy and character-poor end up being superficial and uninvolving. The kinds of characters you create determine everything about your story, even its direction and outcome Without an understanding of character and psychology, a writer can't create plausible (or implausible) plots and certainly can't have as much fun.

And that's because figuring out how people think and act is probably the most fun writers have when they work. Writers who get into their characters find that they get lots of "help" with their stories and scripts from those characters. It's almost a mystical thing. Somehow, characters we create end up giving us (through dialogue and reactions) ideas we never imagined and plot twists that come as complete surprises to us. It may be a cliché but somehow, magically, characters we create do seem to take on minds of their own and pull us in new directions. The trick is learning how to trust our creative process enough to give into the exploration of these directions, while at the same time exerting enough control to be true to the original concepts of our stories.

That means we have to create characters that mean something to us—that we can grow to love and "trust" just as if they were "real" people. They must in fact, become "real" to us, and in order for them to do so, we have to know as much as possible about them—the way they talk, think, and move; their hopes, their fears, and most of all, the details of their lives. The actual story of a film is really a picture of our characters' present and immediate future (even though sometimes we're told the story of their pasts). Audiences must believe that characters in a movie had a full life before the film began and will continue to live long after the movie ends.

You can see that principal operating most effectively in films that have given rise to sequels. Characters like Indiana Jones, Darth Vader, Luke Skywalker, James Bond, Harry Potter, and many more sequel darlings are so compelling and evocative that audiences want to see them again and again. And these characters are so interesting because of the proclivities and quirks that were developed for them as part of their biographies. In fact, some characters are so fascinating (and because of that, potential big money makers), that their biographies, not even present in the films in which they first appeared, give rise to prequels. (*Star Wars* and the Hannibal Lechter series are marked examples of this.)

That's why you should write *character biographies* for all your main characters and most of your minor characters. All the characters in screenplays are important. Even minor characters should be given serious consideration because their presence will form a subtext and atmosphere within which main characters can act. It's important, therefore, to create short character biographies for some of the most "prominent" minor characters in each screenplay. You might want to do that after you've created biographies for your main characters, because in the creation of those biographies you might discover minor characters you never knew existed and plot possibilities you never knew you had.

You should begin your character biographies in the past even though things you include in these biographies might never even appear in the film you write. And you should write them with more than facts in mind. You should include images and events that make up your character's memory. If you do that, you'll find that you have a richer storehouse of character qualities from which to draw.

Here's a sample paragraph from a biography of a main character that veers away from hard facts (like date and place of birth) and explores images and events.

Character Biography of Laura Turner:

As the doll house burned to the ground, Laura stood on one foot and watched. She hated standing on one foot but it was so uncomfortable that it made her feel less guilty about setting the fire. It was a small penance for what she considered a just crime. Marvel Carruthers deserved what she got for bragging about how her perfect father made that perfect house for her in his perfect workshop. Laura wished that burning the doll house would make Marvel Carruthers stop bragging forever about her good father but Laura knew that would never happen, even if she set fire to the whole world.

You can see from that little paragraph how I set up some basic character elements: a character who was jealous and vengeful; who had twinges of conscience; who would punish herself but not too much; who had a father whom she did not consider "good," and an unhappy home life; who felt she was doomed to failure no matter how hard she tried.

A little depressing? Don't worry. In subsequent paragraphs I might hint at the possibility of a character arc by creating events at which she was successful and times in which she was happy.

This biography process is important because it makes us think carefully about the strengths and weakness of our characters—strengths and weaknesses that will be important in determining the story's outcome and defining the struggle between good and evil, between protagonists and antagonists. Incidentally, I'm not a fan of these very theatrical terms. Instead, as a concession to the Adelphi theater, I'll call them Good Guys and Bad Guys or, as they were portrayed in Westerns, White Hats and Black Hats.

PART IV
WHITE HATS, BLACK HATS

WHITE HATS, BLACK HATS

T here's a café on the main square in the town of Aix-en-Provence where you can sit under the plane trees and look across the cobbled street at a row of fashionable centuries-old townhouses. The summers are very, very hot in Aix and the air shimmers. Usually, you are dripping wet a moment after you set foot outdoors but breathing the sticky sweet air of Southern France and sitting at the little café in the shade and sipping something cool, you tend to forget the heat as you enjoy watching the people around you.

The summer I visited that café, I saw a car pull up and let out a thin dark woman with a gamine haircut. She was about 35, but she wore a short yellow skirt and a sheer white blouse with the aplomb of a teenager. She almost fell out of the car, catching her skirt hem on the cracked leather of the car seat and, once on the sidewalk, pulled three large suitcases out of the car's interior. As the car drove off, she dragged the suitcases over to one of the café's curbside tables and flopped down in one of the rickety chairs.

She called for a waiter, ordered something, and then slouched down, tilting her head back and resting it on the back of the chair. She looked impossibly uncomfortable but in a few minutes, she was fast asleep, her mouth open, her jaw slack, her arms dangling by her sides. She looked almost dead. The waiter came and, without disturbing her, put down a small bottle of mineral water and a glass filled with ice and crept away.

As I sat there watching her sleep, I made some careful observations. Her hair was well cut and her clothes were, at first glance, fashionable. But as I looked closer, I noticed that the colors were faded from too much washing, that her hair was dirty, that her lipstick was smeared, that her teeth were ugly. I saw that her shoes were scuffed, that her nail polish was chipped, that her suitcases were old and worn.

I made up all kinds of stories about her: she was a spy exhausted from eluding killers; a detective following someone and pretending to sleep; an ex-nun run away from the convent; a housewife in despair; a gambler down on her luck; a cabaret singer between gigs; a hooker waiting for her pimp.

The stories seemed obvious based on the circumstances her appearance suggested. But when I considered her behavior—the ultimate clue to her character—they gained substance. I marveled at the ease with which she slept in that very public place; at her indifference to the attitude of fellow café goers; at the joie de vivre and relaxation that her sleeping seemed to imply. But most of all, I was amazed by her trust of complete strangers—by her belief that she would remain unharmed, that her suitcases would not be stolen, that she would not be bothered or evicted from the café. And it was that naivete and open childlike abandon that made her unusual and endeared her to me.

The stories I'd made up took on a definite tone: The spy was ingenious and swashbuckling; the detective was incredibly inept; the ex-nun was vivacious and intense; the housewife was relieved; the gambler was nonchalant about her losses; the cabaret singer was confident and conceited; the hooker was an Irma La Douce. Romantic Comedy versus Drama.

I was alone in Aix on a beautiful weekday afternoon with nowhere to go so I stayed in the café writing and thinking, and when I left 2 hours later, the woman was still sleeping, peacefully. I suppose that scene was as much a testament to the atmosphere in small Southern French towns as it was to the character of the woman involved. And it demonstrates how a character's behavior is the inspiration for story.

That demonstrates what I've already said, that story comes out of character. And in the same way, the value system and ethical atmosphere of your screenplay is dictated by the values and ethics of your characters. Even though characters may not talk about their values and ethics, they are evident nonetheless in behavior. And just as we can tell what people are like by their actions (which do speak louder than words), actions usually serve to tell us up front if characters are heroes or villains.

In simpler times, characters in Westerns actually wore stereotypical outfits to indicate their intentions and qualities. Bad Guys wore capes (very Darth Vader and Snively Whiplash), moustaches, and black hats, dressed in black, and moved furtively. Good Guys wore white hats, light colors (or light grey in black and white films), and always stood tall. Slutty women were dressed provocatively—low-cut dresses, high-heeled shoes, bright colors; and virtuous women wore pastels, high necklines, and lots of bows.

We don't buy those corny stereotypes anymore and that's a good thing. But although audiences won't buy corny stereotypes, sometimes they do buy dangerous discriminatory stereotypes that make negative comments about ethnic groups, minorities, or people who they think are "different."

Stereotyping is unethical because it assigns generalities (usually negative) to groups of individuals. Modern writers and studios like to think that they stay away from stereotyping but, unfortunately, they sometimes unconsciously fall into that practice. I have my students do the following exercise to alert them to their own subconscious tendencies to stereotype.

Exercise

Write physical descriptions of people engaged in the following professions and include a brief description of their behavior at an event (wedding, sporting event, etc.). Do not include dialogue. Use only description.

- Piano tuner
- Morgue technician
- Construction worker
- Accountant
- Poet

You can add any other professions you think are interesting.

Now look at your descriptions and decide if they are stereotypes of the profession. For example, people tend to portray piano tuners as mousy, quiet guys with failing sight, antisocial habits, and quirky lives on the poverty line. I know quite a few piano tuners and none of them fit that description!

And the same can be said for lots of the other professions. Not all construction workers, for example, are brawny hard-hat hunks with sawdust for brains. Nor are poets all ethereal idealists. So consider how you might change your descriptions to create more original characters who happen to practice unoriginal professions.

You might also perform this exercise in connection with social constructs. For example, I had a group of my beginning screenwriting students describe a poor person sitting on the porch of a house. I was astounded to see that most of the people said that a poor person would be wearing rags and sitting on the stoop of an ugly tenement. Unfortunately, this response might be a function of the economic status of my students. Maybe they were all rich kids, because anyone who is or has been poor will tell you that poor people do not wear rags, and have a pride that inspires them to keep themselves and their environments as neat and clean as possible. Following that same line of thinking, not all sorority girls are snobs, not all beautiful women are bimbo airheads, not all bosses are heartless.

Unfortunately, just as we tend to stereotype professions and social strata, we tend to stereotype certain types of behavior by generalizing and categorizing them as good or bad.

What we must think about is whether certain behaviors are innately good or bad or if their categorizations change with circumstances. And to do that, we've got to decide for ourselves whether "good" and "bad" are absolutes or relative terms.

Ultimately, as writers and as individuals, we must have a clear idea of the difference between good and bad. Each one of us must be committed to our own belief of what is bottom-line good—so strongly committed to that belief that our conscience does not let us shift our definition according to circumstance. Only in this way can we develop a clear direction for our lives as well as for our screenplays.

If, for example, we believe that stealing is wrong, should it be wrong under all circumstances? In this case, we can apply the categorical imperative to make a decision. But ultimately we must realize that life and art is a series of serious choices that we must constantly make. It is these choices that provide patterns for our life and for our screenplays.

As a case study, let's take the idea of stealing. I asked students in a beginning screenwriting class if there were any circumstances under which it might be okay to steal. Wild discussion ensued. Most people felt it was okay to steal from "Big Corporations" because that kind of crime seemed victimless. "So do you think that bank robbery should be legalized?" I asked. They didn't think so. But some still felt it was okay to steal from people who had more than they did. "If a guy has five fancy cars," said one student, "it's okay for someone to take one of them!" To their credit, many class members balked at that rationale.

Most class members thought it wasn't really stealing to take items from restaurants, hotels, conglomerate institutions, and the government. On the other hand, all the students said they wanted to live in safe neighborhoods and in crime-free societies. In that case, I asked them, didn't each one of them have the personal responsibility of doing nothing to contribute to danger or crime? Many of them said they probably did have that responsibility but even so, when I took a vote, 8 out of 12 people in the class said stealing was okay in some circumstances.

These students were members of a demographic usually considered "privileged" and in the upper 10% of the population. That so many of them would be willing to create a sliding scale with regard to the Eighth Commandment was very troubling to me. And it certainly broke the stereotype of how students at an expensive, private, religious-based liberal arts university think about right and wrong.

Having a sliding scale with regard to ethics, morality, and values does tend to have a profound influence on the way lives are lived and, in art, how characters are portrayed. If a writer feels that it's okay to steal sometimes, then thieves can become heroes and ultimately Good Guys. (And by "Guys" I also mean women!)

And in fact, sometimes, genuinely good people develop curiosities and fascinations with sliding-scale behavior because toeing the line of staunch commitment to ideals has been stereotyped by their peers and by media as dull and limiting. (We talk more about this in a later chapter!) Some people and lots of writers do in fact become bored with goodness because they believe general stereotypes of goodness. Often, Good Guys are portrayed as inactive marks whose function it is to be victimized by the Bad Guys and then to simply put right the evil that disrupts a banal peace.

With this kind of boring Good Guy stereotype, no wonder those of us who crave diversion and excitement become perversely enamored with Bad Guys. At least their stereotype is thrilling and interesting because of the dastardly deeds and outrageous acts Bad Guys usually commit.

And, usually, Good Guys who do fight evil seem to have no choice to do otherwise. There seems to be no struggle involved. Good characters seem to do good because that's who they are. But that's far from the truth. In real life, it's hard to be good. Being "good" involves a series of choices, often difficult and complicated choices, and the more writers explore the struggle to make those choices, the more interesting and vital the good characters in the screenplay will be.

The struggle a character goes through to make a choice is driven by the nature of that character. Sometimes it seems that a character's nature entirely dictates the choice he or she makes. It might even seem that because of his or her nature, a character has no choice or is compelled by circumstance to perform certain acts. Even so, it's important to remember that a character (just like a real person) *always* has a choice. The choice might be more difficult or even impossible, but it's still a choice. The screenwriter's job is to make that clear to the audience and in so doing (particularly if ethics are a concern) actually demonstrate the difference between choosing good and choosing evil.

Doing that requires screenwriters to examine their own views of good and evil. That examination is a profound one. It is the stuff of literature, philosophy, and religion, the basis for an analysis of all human behavior, and a process no serious screenwriter (or for that matter no ethical person) should avoid.

GOOD

ots of books have been written about heroes. Philosophers and theorists from Plato to Joseph Campbell have put forward what it means to be heroic. They've talked about the hero's task, the hero's struggle, the hero's nature. Sometimes it seems as if there are as many different kinds of heroes as there are situations. For our purposes, let's simplify things and talk instead about *Good Guys* (remember—when I say guys I'm also talking about women).

Keeping it simple, we'd like to believe that the protagonists of our ethical movies are good—that right makes might, that virtue triumphs. What characteristics do Good Guys have? That's easy—characteristics that make them loveable, desirable, and just plain wonderful to be around! Audiences love to spend time with them, enjoy watching them, and care what happens to them. Good Guys are people we'd like as friends. They're people we'd want as family members. And usually (at least the way Hollywood casts them) they are people we'd choose to love and marry.

Let's get specific and make a list of Good Guy characteristics. Begin by closing your eyes and picturing your perfect mate. Now list the characteristics you want that person to have. Look at the list. All the qualities you've put down are probably what you consider "good" and give rise to "good" actions. I doubt that anyone has written down "serial killer" or "compulsive liar" as a description of the perfect mate!

You might be interested to know that your list probably isn't too far off from good qualities listed as the divine soul qualities attributed to Good Guys in ancient times. The Bible, for example, is filled with descriptions of what it means to be good but these are too randomly placed to be definitive. The Talmud and the Koran also stipulate what being good means but again, they take a long time to do that. Fortunately, the Bhagavad Gita (part of the sixth book of the Mahabharata, a sacred Indian scripture written somewhere between 400 B.C. and A.D. 400) concisely lists soul qualities that make people "God-Like."

131

I've decided to use the list as it was translated by Sir Edwin Arnold because it is at the same time explicit and poetic. These Divine traits are:

> *Fearlessness, singleness of soul, the will*
> *Always to strive for wisdom; opened hand*
> *And governed appetites; and piety,*
> *And love of lonely study; humbleness,*
> *Uprightness, heed to injure nought which lives,*
> *Truthfulness, slowness unto wrath, a mind*
> *That lightly letteth go what others prize;*
> *And equanimity, and charity*
> *Which spieth no man's faults; and tenderness*
> *Towards all that suffer; a contented heart,*
> *Fluttered by no desires; a bearing mild,*
> *Modest, and grave, with manhood nobly mixed,*
> *With patience, fortitude, and purity;*
> *An unrevengeful spirit, never given*
> *To rate itself too high.*[53]

Wow. Quite a list! Remember that the Bhagavad-Gita says those qualities are possessed by saints and they can also be considered heroic not only because they give rise to "good" actions but also because it takes some amount of courage to practice and display them, particularly in stressful situations.

Compare that to your list and then take a step back. Do you think that any living human being can live up to your idea of perfection? Can any of us have all the qualities you've listed? Do you yourself have all the qualities you've listed?

The fantasy is that somewhere out there the perfect person exists. And it's just that—a fantasy. Most of us tend to believe that finding a perfect person will perfect us. We're usually devastated when we find imperfection in others because we think it takes away from the imagined perfection in ourselves. (This is one big reason for the astronomically high divorce rate in our country.)

The truth is that as human beings who aren't saints, we're all imperfect, whether we like it or not. And yet, we still want to believe that perfection is possible. That's why so many of us cling to the unreal images of celebrities and movie icons. In the old days, the Hollywood publicity machines wanted us to believe that stars really were perfect and acted perfectly. Affairs, scandals, drug use, pederasty, theft, and even murder

---

[53]Arnold, Sir Edwin. (Trans.). *The Song Celestial or Bhagavad-Gita.* David McKay Company, Philadelphia, 1934, p. 112.

were covered up by secretive studios and a cooperative press in the name of the "good" image. Studios thought people needed to believe their movie stars were moral, upstanding, virtuous, and even angelic—superhumans who were better than all of us, and they went out of their way to pretend it was so.

The Hays Code we described in an earlier chapter was put in place so that movies would provide icons of perfect behavior. We've already seen how repressive and censorial it was, and after reading it you might have noticed how it insidiously encouraged Hollywood to create squeaky-clean characters whose examples you had to follow if you wanted to be good in America. And being good in America meant doing the "right" thing always—defending the weak, marrying before having sex and children, dating with honorable intentions, punishing immorality, and cheering the defeat of the Bad Guys who never ever won.

In the days of the Hays Code, films still had their share of sex and violence (albeit more subtly presented) but they always ended with a moral. I hear old folks wishing movies would be like they used to. I don't think any of us really wants that. Many of the old films were exercises in suppression and the Hollywood publicity machine was hypocrisy itself, judging by the tell-all biographies published about those old-time stars. And yet there was something likeable in the innocence and bravado, the profound sense of goodness that inspired old-time movie audiences.

Take Westerns. The Western film defined a heroic code of behavior for America that came out of the need to tame wild open spaces, to give a semblance of civilization to pioneer culture. There was great romance in the act of subduing the forces of nature, riding the range, creating homesteads, and combating the evil forces trying to thwart good, civilizing influences.

Cowboy stars like Roy Rogers, Gene Autry, the Lone Ranger, Hopalong Cassidy, and Wild Bill Hickok were considered admirable role models (especially for kids)—role models that went far beyond screentime and into the real life of audiences. These Western stars went so far as to maintain high personal standards as they endorsed specific codes of behavior for their fans.

William Boyd, who played Hopalong Cassidy (a character based on 28 novels by Clarence Mulford), took his role so seriously that every time he put on the black Hoppy costume, he actually became what he determined would be a symbol for ultimate goodness. Mulford created Hopalong as a much meaner and unsavory character, but Boyd transformed that creation into a icon of the Good Guy.

As *Time* magazine wrote in 1950, "Boyd made Hoppy a veritable Galahad of the Range—a soft spoken paragon who did not smoke, drink, or kiss girls, who tried to capture rustlers instead of shooting them and who always let the villain draw first if gunplay was inevitable."[54]

Boyd felt so strongly about his role modeling that in the 1940s he published the *Hopalong Cassidy Creed for American Boys and Girls*. It gave his young fans 10 dictums for clean living and if you wanted to be a friend of Hoppy's you'd follow it to the letter. Here's what it said:

1. The highest badge of honor a person can wear is honesty. Be truthful at all times.
2. Your parents are the best friends you have. Listen to them and obey their instructions.
3. If you want to be respected, you must respect others. Show good manners in every way.
4. Only through hard work and study can you succeed. Don't be lazy.
5. Your good deeds always come to light. So don't boast or be a show-off.
6. If you waste your time or money today, you'll regret it tomorrow. Practice thrift in all ways.
7. Many animals are good and loyal companions. Be friendly and kind to them.
8. A strong, healthy body is a precious gift. Be neat and clean.
9. Our country's laws are made for your protection. Observe them carefully.
10. Children in many foreign lands are less fortunate than you. Be glad you're an American.

In fact, this creed was really the blueprint for the value system of the Hopalong character, and all 66 of the Hoppy movies involved the creed in their messages. Boyd said "I played down the violence and tried to make Hoppy an admirable character and insisted on grammatical English."

Audiences took to Boyd and his code, and their admiration for him lasted well into the 1950s, when Hopalong Cassidy got a television show and a column syndicated in newspapers where he wrote letters on responsible gun use and good citizenry for young folks. Boyd was so popular that he attracted huge crowds as he went around North America inspiring little cowpokes by wearing the Hoppy costume and championing high ideals.

---

[54]Official Hopalong Cassidy Website: www.hopalong.com

I was one of those little cowpokes. One blistering July day I made my mom stand in line with me at a county fair for 4 hours just so I could shake Hoppy's hand. I was only about 7 but I still remember the moment. He wore black gloves and I was awestruck by his gigantic size and mesmerized by his all-black outfit and silver-studded gun belt. Our moment together was magical because I believed in the ultimate goodness of Hopalong.

Other cowboys like Roy Rogers and Guy Madison had the same appeal. Guy Madison, the hero of the *Adventures of Wild Bill Hickok*, a TV show that ran from 1951 to 1959, adopted a code like Hoppy's and attached it to the end of his show. He called it his *Code for Little Deputy Marshals*:

1. I will be brave but never careless.
2. I will obey my parents. They DO know best.
3. I will be neat and clean at all times.
4. I will be polite and courteous.
5. I will protect the weak and help them.
6. I will study hard.
7. I will be kind to animals.
8. I will respect my flag and country.
9. I will attend my place of worship regularly.[55]

The fact that Madison added religion to his code wasn't controversial in those days. In fact, it was generally believed that Good Guys (especially cowboys) believed in God. Roy Rogers (the King of the Cowboys) always ended his shows by saying "May the Good Lord Take a Liking to Ya." Even today, Roy Rogers and Dale Evans are still famous for their value systems, which promoted clean living, integrity, and a belief in God and country. The fact that the religion was ecumenical made it inclusive and very palatable.

Behavior codes are unrealistic today, but then they operated on the same principal as celebrity endorsement. The reasoning was that if celebrities could sell products to kids, why couldn't they sell "goodness." Nothing wrong with that! And yet can you imagine kids today buying into this kind of "code," considering the lives we see today's stars living? Kids (and adults, too) detest hypocrisy. It doesn't make sense to have a movie or TV star endorse positive values if that star appears in films eschewing those values, and most certainly if that star lives a life that is valueless. Today's entertainment journalists

---

[55]Official Guy Madison Website: www.guymadison.com

don't care about star and studio "good" images. They report every-thing. Maybe that's why it's so difficult to find genuine Good Guys both on and off the screen.

Our notion of what is good has been diluted by the reality of urban blight and modern hard-core living. Films are more claustrophobic and less altruistic in their ethical and moral value systems. Sophisti-cated audiences no longer buy into perfect people as heroes. Nor do they buy into perfect goodness.

These days, Good Guys (protagonists) have to have serious flaws for audiences to believe in them. Unfortunately, some of these flaws are so pronounced that the definition of goodness has become so skewed as to sometimes blur into evil. Movies have gangsters, hit men, Mafia bosses, cruel killers, thieves, and other reprobates as heroes, and audiences are expected to root for them because they aren't quite as bad as the villains who oppose them. It's not unusual for audiences to find themselves cheering for a thief or even a serial killer to win, forgetting (or ignoring) the crimes committed.

So how do we create Good Guy characters who are not so good they're unrealistic, but not so flawed that they stop being good? It's not easy. Getting back to the "good" stereotype, if we were to examine closely the seeming "inactivity" of "good" characters, we would find that their goodness is really the result of internalized action—a com-plex and often disturbing process of decision-making that causes them to choose goodness often at great cost. What makes Good Guys dynamic is the interior conflict they experience in deciding to be good. That tension, which is difficult to write well, makes Good Guys pro-foundly interesting and has potential to create intense drama, sus-pense, and excitement in the story process.

Films that show "good" characters making difficult choices are much more complex, layered, satisfying, and original than films that show "bad" characters blowing people away only because they are compelled by greed, ambition, cruelty, or cowardice. The action of "good" characters making difficult choices often tends to be more origi-nal than what Hannah Arendt called "the banality of evil."

As an example of a "Good Guy" making difficult choices to stay good, let's take another look at *The Insider*, the film we used as an ex-ample of the Utilitarian model of decision making. As you remember, *The Insider* is based on the true story of Jeffrey Wigand, a PhD re-search biologist who takes a stand for public health and against his former employer, a big tobacco company. (Wigand has a Web site and you can even chat with him on-line.) As we examine the ethical choices made by Jeffrey Wigand, we can see how each one of them created dy-namic tension and continually upped the ante of the plot.

Jeffrey Wigand starts off clearly a good guy. In the first few minutes of the film, we're told that he's a solid family man. He is attentive to his small daughter and loving to his wife and he is very responsible and effective in dealing with his older daughter's health problems. When she has an asthma attack, he's the one who calms her and gets her treatment going.

But we're also shown some of his flaws—he drinks, he broods, and he isn't open with his wife. We know that he's been fired but we're shown that he waits to tell his wife until she presses him and then he avoids talking about it with her. He doesn't tell her why and we don't find out why until much later in the movie. So far, Wigand seems like just an ordinary guy with character flaws.

But then fate puts him in the position of getting the chance to "work" as a consultant for *60 Minutes*—a chance that he's motivated to take because he's concerned about money. He begins to make a series of difficult choices to do good and protect the public health or to do nothing and protect himself. Ultimately, he chooses to tell what he knows and to become a whistle blower on the cigarette companies.

In the rest of the film we watch as he is systematically discredited and destroyed, as CBS refuses to back him, as Lowell Bergman resigns his job because he has lost confidence in the network process. And in the end, as the *60 Minutes* interview finally airs, we see Wigand watching it with his daughters and smiling, we get the idea that in spite of everything, because of his commitment to truth and ethics, and because he is truly a Good Guy, Jeffrey Wigand feels what he went through was worth it all.

Jeffrey Wigand is a character who fights to be good and struggles mightily with the urge to give in to his "bad" side. He resists the temptation to give into his "base" desires for money, position, health, prestige, and even a family life in order to honor his higher desire to do good for a public cause. He wins his fight with temptation at great cost, but sometimes Good Guys lose that fight. Sometimes they give in to their "evil" tendencies and make choices that propel them down a slippery slope to actually becoming evil. Even in cases where the "fight" to be good is subtle, a description of that process can create remarkable tension between characters and in the plot itself.

That's made clear in two movie examples: *Amadeus* (1984) and *A Simple Plan* (1998). In Amadeus, the "evil" is subtle and insidious and the struggle against it refined. In *A Simple Plan* the evil is overt and drastic. And yet both films demonstrate how men who start out good make choices that change them and "make" them evil.

Movie Example #1

Amadeus (1984)
Written by: Peter Shaffer
Directed by: Milos Forman

This Academy Award-winning film was primarily about Salieri, a good, religious, and talented composer, and his terrible obsession with the profoundly gifted genius, Wolfgang Amadeus Mozart.

Salieri starts off as a "Good Guy." As a boy, he prays to be allowed to "celebrate" God with music. He makes a bargain with God—if he's allowed to become a famous musician, he promises to be chaste, industrious, and humble. God grants him his wish. Salieri becomes a court composer and writes celebrated works he believes are sublime and wonderful. He is a man who considers himself blessed and pious.

And then he meets Mozart and realizes that Mozart's talent far exceeds his own. Salieri becomes jealous of Mozart but most of all, he is angry with God for "mocking" him by bestowing on "a vulgar creature" such magnificent gifts while saddling Salieri with a puny talent. "All I ever wanted was to sing to God and then he made me mute," says Salieri. "Why plant the desire and then deny the talent?"

Salieri's rage motivates him to make the choice to become God's enemy and to spite Him by destroying Mozart. Salieri makes this commitment to evil in a very dramatic scene where he denounces God by burning a crucifix. In successive scenes, Shafer (the writer) skillfully demonstrates Salieri's descent into evil by choosing evil actions and sticking to them in spite of some resistance from his conscience. Here's a list of some of the evil things Salieri chooses to do:

- He gives Mozart left-handed praise and makes things difficult for him at court.

- He contrives to deny him a much-needed teaching job by humiliating him into "auditioning."

- He hires a maid to snoop in the Mozart household and goes on with his persecution even though she warns him that Mozart cannot take much more suffering.

- He undermines Mozart in the presentation of a new opera while pretending to help him.

- On one hand, because he truly admires Mozart's work, Salieri gives loving friendly advice and encouragement but cannot resist betraying him to the Emperor.

     ◈ He disguises himself in a way that he knows will remind Mozart of his demanding father and commissions a Mass with the distinct purpose of playing on Mozart's weakness and driving him mad.

     ◈ He decides to find a way to kill Mozart and to pass that Mass off as his own.

"I will force God to listen," says Salieri. "I will laugh at God." And to do that he uses Mozart, who by this time has become only a means to an end. Even as Mozart lies dying and Salieri acknowledges to him that his work is the greatest ever composed, Salieri still can't resist sacrificing Mozart to his own ego and greed. By then, the evil inside of him has become so strong that he cannot possibly resist it.

Salieri urges Mozart to finish the Mass, knowing that the exertion in doing so will probably kill him. Salieri takes dictation from Mozart in a kind of collaboration that for the first time allows Salieri to become a part of that magical creative process he has coveted for so long. As Mozart dictates, it's as if he is a channel, a medium for God to finally speak to Salieri in the language of music. Salieri is transported by that communication—the singular and unique thrill of inspired creativity that had always been denied him.

And yet, before Mozart dies, he still proves himself to be the better man by asking Salieri's forgiveness. In the end, Salieri maintains that God killed Mozart to torture Salieri—to prevent him from sharing in the credit for the Mass, to deny him future glimpses into Mozart's paradise of divine creativity, and to banish him forever to the hell of mediocrity from which he came.

Through the choices he makes, Salieri becomes evil and the villain of the story, a man incapable of winning the struggle against the negative forces of his own character. By showing that struggle and its failure, by showing Salieri making the choices that would finally destroy him, Shafer succeeds in creating an evil character with remarkable scope and depth.

Movie Example #2

A Simple Plan (1998)
Written by: Scott B. Smith
Directed by: Sam Rami

A Simple Plan is a wonderful little film about a Good Guy and what happens to him when he finds a lot of money.

Hank Mitchell (played by Bill Paxton) lives in a friendly small town where he's got a good job at a mill, a pregnant wife (Sarah, played by

Bridget Fonda), and a weird loser brother. Hank, his brother Jacob (played by Billy Bob Thornton), and lowlife buddy Lou find a downed plane with four and a half million dollars inside. Hank's first impulse is to turn the cash in because he believes that "You work for the American dream. You don't steal it."

Convinced to keep the money by Jacob and Lou, Hank insists he should be the one to hold onto it until someone finds the plane. If no mention is made of the cash, they'll keep it. Things get more and more complicated as Hank's desire for the money forces him to make increasingly unethical choices until finally he becomes entirely evil.

The film is a powerful demonstration of how nonethical decisions can be rationalized in the interests of fulfilling profound desires and how good guys can be corrupted by their own character flaws. Hank gives into greed because he wants his family to have a better life. He rationalizes that the money belongs to drug dealers and that no one will miss it.

He doesn't really mean to hurt people and to do evil things. But in fact, he does become evil because his choices make him take evil actions. Once taken, these actions cannot be revoked in spite of the rationale for their commission. At the end of the film, Hank is evil. We'll take a closer look at Hank's choices a little later.

Exercise

Make a list of films in which characters start off good and then make choices that turn them "bad" or evil.

As evidenced by your list, there are lots of other films that show how a Good Guy makes choices that turn him into a Bad Guy. It's harder to find a Good Guy who stays good by making good choices. Films like that tend to lose momentum after the first choice because once that's made for good there may not be much else to say. Those kinds of films only work when the choices are ambiguous or continuously problematic.

Oskar Schindler of *Schindler's List* is an example of a someone who starts out as something of a Bad Guy motivated by greed and then gradually makes good choices in order to save Jews. At the end of the film, almost in spite of himself, Schindler becomes a Good Guy.

If you can describe the difficulty good characters have in making choices, the process by which they make choices, and the tension that occurs in their lives as a result of these choices, then "Goodness" will not seem boring or banal. Audiences will relate to heroes who are challenged because they aren't perfect but continue to strive for perfection.

Heroic characters should, in fact, have character flaws that create struggle and motivation on the part of that character to overcome the flaw. In creating this kind of human hero, and describing the struggle to be good, a screenwriter can deliver a message to audiences about what goodness really means.

In *Schindler's List*, we saw in Oskar Schindler a movie hero (based on a real person) who was greedy, egocentric, and cold, but who struggled to overcome these character flaws and do good. The writer's message: Sometimes it takes extraordinary effort by ordinary people moved by their own humanity to overcome evil.

And what about evil itself? Let's examine it and then, at the end of this section, I give you some practical techniques for writing both good and evil characters.

BAD

O n September 11, 2001, 7 a.m., L.A., I was feeding my big red Chow dog, Riley, when the phone rang. At that hour, I thought it must be my husband calling from Japan where he was traveling for business. Instead, it was my sister calling from Manhattan. She was nearly hysterical. "The States *is* under attack! They've hit the Pentagon and the World Trade Towers! It's the Apocalypse," she cried, her voice filled with terror. A few minutes later, I turned on TV and saw the inconceivable images of passenger planes crashing into the World Trade Towers. I was sickened and devastated. And for the next week, along with the rest of the country, I was consumed by the disaster and focused on the continuous reportage.

By the end of the week I noticed that this addiction to the news and my preoccupation with the tragedy was making me depressed, anxious, fearful, and restless. My sleep was interrupted, I was nauseated lots of the time, and I felt exhausted.

I also noticed that when my attention shifted to stories about heroism and goodness surrounding the events, my anxiety was greatly reduced and I felt comforted and inspired. More and more, I concentrated on hearing stories about the heroic firemen and police, the angelic helpers who sacrificed so much physically and emotionally to rescue others. And each time I thought about the goodness of people, my spirits lifted.

Eventually, I forced myself to stop watching or listening to any more news coverage. Although the events of September 11 stayed in my mind, a week after I stopped replaying the horrible images in my imagination, even though a deep sadness set in, I was no longer anxious or overwhelmed. This was a dramatic practical example to me of how damaging a preoccupation with violent images can be and how much power images of evil have to destroy and darken lives. It seems to me that it should be obvious that audiences who become addicted to images of violence and mayhem (in the news or in movies) and who make a habit of watching these images, will have their psyches profoundly affected.

I think that's because evil is so powerful it can generate intense emotions and reactions and create compelling fascinations that can be addictive. I believe this power and ability comes from the fact that evil provides such a marked contrast with good that it causes us all to consider our own particular proclivities and to measure our strength against them.

Strangely, in this way, evil does provide a kind of service to mankind. As Helen Keller wrote in her inspiring little book *My Religion*, evil people "enable man to see the evil he is to avoid as well as the good he is to choose. They keep alive the fires of ambition in him when he does not care about ideals or the public welfare but desires rather fame and honor. They sharpen some minds for unpleasant truths which the children of light must surely learn if they are to help guard humanity against brute force and every form of oppression whether it be by one or by many."[56]

Helen Keller, the deaf and blind woman whose story was so powerfully portrayed in *The Miracle Worker* (the original 1962 film with Anne Bancroft as teacher Anne Sullivan and Patty Duke as Helen Keller, is the one to see!) was a follower of the Swedish 17th century theologian Swedenborg, who subscribed to the belief that there is no such thing as predestination to hell and that "all are born for heaven."

In Keller's words, Swedenborg said:

> There is no hell of the mediaeval kind; but there is a mental hell into which people go who are self-confirmed lovers of evil and who wilfully deny God in their heart. They do not fall into literal fire and as they punish themselves more than enough, God takes away from them even the anguish of conscience.... They burn with selfish instincts and love of dominion...They see as they think. They debate and litigate and fight; they practice endless arts of magic and "faking," they must labor hard for air and food and some of them seem always to be cutting wood and mowing grass because on earth they worked so furiously for rewards. Misers hug to their hearts imaginary money-bags. Sirens try painfully to beautify their pitiful forms and enjoy their images reflected in the dull light as of a charcoal fire. Each gang of crooks strives to outwit all the rest and the fierce joy of rivalry shines luridly on their marred faces."[57]

---

[56]Keller, Helen. *My Religion*. Pyramid Books, New York, 1974. p. 80.
[57]Ibid., p. 78.

When you consider Swedenborg's philosophy, when you read in Buddhist teachings that "people make a distinction between good and evil, but good and evil do not exist separately"[58]; and in the Bhagavad-Gita, that there is "One Life, One Essence in the Evil and the Good"[59]; if you determine with theologians, metaphysicians, poets, and yogis that everything is of God, then logic seems to indicate that evil (a rejection of good) cannot really exist without good—that evil and good dwell in a symbiotic relationship and perhaps are even made up of the same essence. As Emerson wrote in *Compensation*, "The universe is represented in every one of its particles. Everything is made of one hidden stuff. The world globes itself in a drop of dew...The true doctrine of omnipresence is that God appears with all His parts in every moss and cobweb.... If the good is there, so is the evil."[60]

Can we know what good is if we can't measure it against evil? Can we see light without shadow? Can we call ourselves human if we don't have within ourselves at least a potential to choose evil? And can free will exist if we don't have the opportunity to make that choice? These questions are the stuff of high philosophy and religion. And yet all writers should consider them if they want to create stories and characters that come alive to audiences. That's because stories without contrasts (some would say conflict) are not stories but statements.

Traditionally, all stories are essentially about the struggle between opposing forces of good and evil. Even when the struggle is internal, the issue of evil is always addressed. The traditionally religious example of that struggle is Jesus's temptation in the desert for 40 days and 40 nights. In that case, the tempter (temptation itself) was called "the devil." And in each case, Jesus made the choice to resist temptation, finally saying, "Begone, Satan! For it is written,'You shall worship the Lord your God and him only shall you serve'" (Matthew 4:3–11).[61]

It is this struggle with temptation to do evil that most clearly expresses the truth that humans do have the freedom to choose. And when humans do make a choice, that choice is defined by an action. So what actions are considered evil? Frankly, it's getting harder and harder to tell.

In the movie *K-Pax* (written by Charles Leavitt), the "alien" Prot (played by Kevin Spacey) tells psychiatrist Mark Powell (Jeff Bridges)

---

[58]Hanayama, Shoyu (Ed.) and Steiner, Richard R. (Trans.). *The Teaching of Buddha*, Bukkyo Dendo Kyokai, Tokyo, Japan, 2001, p. 62.

[59]Arnold, Sir Edwin. (Trans.). *The Song Celestial or Bhagavad Gita*, David McKay Company, Philadelphia, 1934, p. 44.

[60]Emerson, Ralph Waldo. *Essays by Ralph Waldo Emerson*, Thomas Y Crowell Company, New York, 1951, p. 73.

[61]*The Holy Bible*, Thomas Nelson and Sons, New York, 1952, p. 759.

that all beings in the universe know the difference between good and evil. I'm not so sure they do. Along with sociopaths and psychopaths, I'm sure there are lots of people who don't seem to have any "rules" or standards by which they determine what distinguishes good from bad.

In Western society, for example, some use the Ten Commandments to define good actions. But many people today do not consider that breaking one or some of the Ten Commandments makes a person "bad." In fact, many apply that sliding scale to many of the Ten Commandments. Let's take a moment to list the Commandments (found in Exodus 20:1–17) and consider the sliding scales that can be applied to rationalize the breaking of each one.

1. You shall have no other gods before me.
    Sliding scale rationalization: There are other, more important things—gods if you will—in life, and after we take care of them, we'll have time for God.
2. You shall not make yourself a graven image.
    Sliding scale rationalization: The icons we worship (fashion, celebrity, material goods) are innocent enough. It's just fun paying all our attention to them.
3. You shall not take the name of the Lord your God in vain.
    Sliding scale rationalization: It's okay to swear and curse. Everyone does it. Language evolves and besides, it's cool.
4. Remember the Sabbath day, to keep it holy.
    Sliding scale rationalization: We work so hard and have so little time to get things done that we need every moment of the weekend to do errands, rest, and party. Also, it depends on what you consider "holy." Sleep, sex, and the samba are holy to some people.
5. Honor your father and your mother.
    Sliding scale rationalization: Parents have obligations toward their children but shouldn't meddle too much in their lives. Children don't necessarily owe their parents anything because kids didn't ask to be born. And besides, lots of parents are too strict or too lenient or abuse their children and so cause them to do the bad things they do. Parents are always to blame for the actions of their kids.
6. You shall not kill.
    Sliding scale rationalization: Killing is horribly wrong except in self-defense, or if you're really really mad, drunk, or stoned, or if the victim deserved it.
7. You shall not commit adultery.
    Sliding scale rationalization: Sometimes, you just can't help

yourself, or your marriage is really awful, or there are extenuating circumstances. Casual sex "doesn't mean anything."

8. You shall not steal.
   Sliding scale rationalization: Stealing is okay if it's from the government, big companies, bad guys, or people who don't deserve to have more than anyone else.

9. You shall not bear false witness against your neighbor.
   Sliding scale rationalization: It's okay to lie if it doesn't hurt anyone, if you can get yourself out of a jam, if you have a good excuse. People expect lying. In some places, it's even part of the culture.

10. You shall not covet your neighbor's house, wife, etcetera, etcetera.
    Sliding scale rationalization: Coveting (jealousy) is good because it motivates people to get what they want. Winners are rewarded. Losers lose. If you can get something away from someone ("your neighbor") then it's a good thing. Winning is usually very meaningful in our society. Coveting is really only competition and that's what makes our society great.

Are sliding scales good? Each one of us does have a different idea about how rigidly we need to adhere to the Ten Commandments. Most people I talk to tell me they believe having a sliding scale for the Ten Commandments is practical and modern. Most believe that all the Commandments don't have equal weight. They think that breaking the "minor" ones doesn't make a person evil or even particularly bad. In fact, many say that the breaking of a "minor" commandment doesn't need to ruffle anyone's conscience.

What's a minor commandment? The Ten Commandments don't seem to be listed in order of their "seriousness" (honoring the Sabbath is the fourth and killing is the sixth); most people feel they can create their own commandment ranking system and they do. They like to shuffle the Ten Commandments around, usually parking "Thou shalt not kill" in the number one spot and then arranging the others to suit their convenience.

If breaking the major Judeo–Christian religious precept of the Ten Commandments or using a sliding scale to rationalize the breaking of any one or all of those commandments doesn't make a person "bad" or "evil," then what kind of behavior does? We might turn to our religious texts to find descriptions of evil. If we go to the Bhagavad-Gita (again because of its concise poetry) we find it actually describes actions it calls "demonic." It doesn't talk directly about breaking laws but instead describes the mindset, lifestyle, and actions of evil people:

Deceitfulness, and arrogance, and pride,
Quickness to anger, harsh and evil speech,
And ignorance, to its own darkness blind,
... They comprehend not, the Unheavenly,
How Souls go forth from Me; nor how they come
Back unto Me; nor is there Truth in these,
Nor purity, nor rule of Life. "This world
Hath not a Law, nor Order, nor a Lord,"
So say they; "nor hath risen up by Cause
Following on Cause, in perfect purposing,
But is none other than a House of Lust.

And, this thing thinking, all those ruined ones—
Of little wit, dark-minded—give themselves
To evil deeds, the curses of their kind,
Surrendered to desires insatiable,
Full of deceitfulness, folly and pride,
In blindness cleaving to their errors, caught
Into the sinful course, they trust this lie
As it were true—this lie which leads to death—
Finding in Pleasure all the good which is,
And crying "Here it finisheth!"

Ensnared in nooses of a hundred idle hopes,
Slaves to their passion and their wrath, they buy
Wealth with base deeds, to glut hot appetites;
... Darkened by ignorance; and so they fall—
Tossed to and fro with projects, tricked, and bound
In net of black delusion, lost in lusts—
... Conceited, fond, stubborn and proud, dead-drunken with the wine
Of wealth and reckless, all their offerings
Have but a show of reverence, being not made
In piety of ancient faith.[62]

The Bhagavad-Gita says evil is driven by desire, anger, and greed—the traditional motives of villains. And although most people would agree that these "demonic" mindsets and acts and the people who practice them are unattractive, the motivations of anger, desire, and greed are not necessarily considered so.

In fact, villains today have become so popular in part because modern times have made desire, anger, and greed justifiable, if not down-

---

[62]Arnold, Sir Edwin. (Trans.). *The Song Celestial or Bhagavad Gita.* David McKay Company, Philadelphia, pp. 112–115.

right attractive. Our hedonistic tendencies are given full encour-
agement by advertisements, entertainments, and social constructs.
Our society sanctions and celebrates gambling, drinking, and free sex.
Immodesty is flaunted by fashion mavens. Business and the pursuit of
profit is encouraged and extolled. Audiences cheered as Gordon
Gekko (played by Michael Douglas) proclaimed in Oliver Stone's *Wall
Street* (1987):

> GEKKO
> ... Greed is good. Greed works, Greed is right.
> Greed clarifies, cuts through and captures the
> essence of the evolutionary spirit. Greed in all its
> forms, greed for life, money, love, knowledge,
> has marked the upward surge of mankind.

I suspect that people would still cheer that speech today, even if only
in sotto voce. If the motivators of evil are not considered evil, then does
evil still exist and can we define it? It's getting harder and harder to do
that especially because villains, like heroes, have "flaws" of goodness
within them. Villains, just like heroes, can't be one dimensional if they
are to be believable, and that's created a special problem for the
screenwriter who wants to define evil for audiences and make that evil
unattractive and something to be shunned.

Because, as we all know, the screen villain has morphed into something else—something otherworldly—something I call the *Villanero*.

THE VILLANERO

hat a cute term! It's one I made up combining the words villain and hero and it's especially apt because these days modern audiences are rooting for bad guys who have so many warm and fuzzy qualities that they actually come off looking like heroes. The term *villanero* is particularly adorable because if you put a space between the two words (Villa Nero) it can be translated as House of Nero, the playground of that horrid Roman emperor who fiddled while Rome burned. And that's what lots of screenwriters seem to be doing in the face of audiences' growing fascination with lives of crime.

My students think villaneros are "cool." There's a long list of memorable villaneros: Freddy Krueger (in the Freddy horror films), Dracula (in all the various vampire movies), Travis Bickle (in *Taxi Driver*), Hannibal Lechter (in *Silence of the Lambs*), Vincent Vega (in *Pulp Fiction*), and many more of that ilk, and they've all made being bad seem kind of fun and glamorous. I'm sure you won't have much trouble making a list of some of your favorite villaneros.

But, as interesting as villaneros are, we've got to be more careful creating them if we want to write more ethical scripts. Blurring the lines between villain and hero in characterizations only makes the distinction between good and evil less exact and can confuse the public into believing that good and evil are relative terms. It's not my belief that they are.

I think that the men who destroyed the World Trade Center were evil. I do not think, as some do, that America "deserved" what happened because of our foreign policy. I believe that there are acts we need to define as evil and that we need to adhere to these definitions. I do not think there should be a sliding scale for certain despicable acts.

A review of the film *L.I.E.*, for example, said that the screenwriter painted a sympathetic picture of a child molester but did not condone or excuse child molestation. Frankly, I think it's unethical to paint a "sympathetic picture" of someone committing such an evil

149

act. Doing that says you can separate the doer from the action and that you can compartmentalize effectively enough to be able to ignore or set aside a person's evil act and concentrate only on that person's good qualities. I'm not so sure that's possible. Can a person commit evil acts and still be good? I do not think so. But it's that kind of philosophy that has lots of screenwriters creating characters who audiences like and whose evil acts audiences "forgive" or excuse because of that likeability.

That kind of thinking is dangerous and erodes morality. It's probably the thinking of those who fall in love with serial killers and marry them in jailhouse ceremonies. I'm certain it was even in operation in Nazi Germany. I'm sure even Hitler had friends and relatives who found him charming and likeable. Hitler's buddies probably did not believe that he was evil. I'm sure if you asked them they'd say "Hey, Hitler was a great guy 'cause he was always really nice to me!" And believe me when I tell you that analogy is not too far-fetched. I've heard it lots of times applied to people who do bad things.

("I just can't believe he's an axe murderer! He was always so polite, quiet and nice to *me*!") People are partial to their own experiences and just don't want to think that people good to them can be nasty to others. And what's most disturbing, they may not even care. It's a well-known fact that, in general, people aren't as connected with or empathetic to those outside of their immediate circle. Remember that well-worn example from journalism? Bad events in your home town get headlines. Bad events in distant locations (that have no relevance to your own life) are buried in the back pages.

It's my personal belief that people who commit evil acts have chosen to be evil, and because of that choice, I have to determine that they side with evil and so they are evil. You can't have it both ways. I do believe people can see the error of their ways and be redeemed but that redemption has to include retribution. Frankly, some people let evildoers off too lightly. Saying "I'm sorry" is not enough. Being likeable is not enough. Evildoers should have to be made to understand that they made a choice and should be made to suffer the consequences of that choice. I believe that too often we excuse evildoers, rationalize their behavior, and forgive them too easily. I don't necessarily believe that we should "hate the deed but not the doer."

I'm not an advocate of hate, but it may not be wrong to hate injustice or to hate evil. And I believe that in order to be truly ethical, writers need to be clear about what evil is, to acknowledge the terrible corruption and soul destruction of evil, and to tell audiences that there's nothing fun, glamorous, or loveable about the people who commit evil acts.

Screenwriters who create villaneros and blur the distinction between good and evil people fall into the trap of making statements that certain crimes by certain kinds of criminals may not be so bad after all, that everyone who commits an evil act isn't necessarily evil, and that there's a sliding scale for evil itself. (Hence the Good, the Bad, and the Blurry.)

In *The Score*, for example, the likeable Robert DeNiro character gets away with robbery and we're glad he does because he's so interesting and has promised to stop his life of crime. And, applying the sliding scale rationalization to theft, he was stealing jewels from decadent and despicably rich people. Even the ultimate object of his last caper—an antique scepter—had a notorious history and belonged to no one in particular. *The Score* followed the rules of caper movies that have likeable thieves getting away with thefts in which nobody but other thieves, unreal entities, large corporations, or despicably rich people are hurt. Studios know that audiences probably wouldn't applaud caper movies based on robberies of widows and orphans. (We discuss *The Score* and other caper films in depth a little later.)

And yet, studios who seem to understand how far they can slide with theft don't seem to know exactly how far they can go with killing. Judging from the numbers of killings we see in films, it seems that even if we're not applauding murder, we don't find the idea of murder evil or shocking. And of course, because the mere idea of murder doesn't shock, writers are forced to make murders more and more brutal to give audiences the jolt they've come to experience.

Ultimately, screenwriters of conscience should ask themselves the following questions:

ᕫᕌ Can evil ever be heroic?

ᕫᕌ Is there ever a good excuse for committing evil?

ᕫᕌ Can evil acts ever be justified? Revenge movies say they can. But as a society are we willing to condone revenge killings that occur outside of the justice system? Our abhorrence of gang executions and gang violence seems to indicate that we aren't in favor of vigilante acts and yet what are our emotions when we see these acts in films? Do we condone them?

Exercise

1. Write down your personal definition of good and of evil.

   Are there clear distinctions between the two? If not, what are the markers that distinguish for you the difference between good and evil acts?

2. Write down what you think are the characteristics of a Good Guy. And of a Bad Guy.

   Do you think good people can commit evil acts and then go back to being good?

3. List some of your favorite Good Guys and Bad Guys.

   Write down the qualities of these characters, why they appeal to you, and why they are effective examples of good or evil.

4. Do you feel that any crimes are justifiable? What are they?

5. Write down your thoughts about the justice system.

   Do you want to perpetuate these beliefs in your scripts?

   What are some ways in which you can create new justice models in your screenplays?

6. List the films that you think paint powerful pictures of good and/or of evil.

   What messages about the nature of good and evil are expressed in these films?

Thinking deeply about this exercise will make you more conscious of the kinds of acts you portray in your scripts, the outcome of these acts, and what you are saying about the characters who commit them.

What decides if characters are good or bad? Usually it's what motivates the characters. If the characters' motivators are malicious or evil, then those characters are Bad Guys. If the characters are simply stupid or the victims of faulty logic and poor judgment, and their motivations are innocent and altruistic, they may be Good Guys.

PRACTICAL WRITING TECHNIQUES

CREATING GOOD GUYS

1. Write down all your main character's good (positive) qualities, inclinations, habits, and tendencies.
2. Create a *Choice Chart*. Referring to your story outline, plot the sequence of choices your hero must make in the course of the film.
   - Write down what motivated each choice to keep in focus why the character is good.

Sample Good Guy Choice Chart: *The Insider*

Here are the choices made by Jeffrey Wigand and the good things that motivated them:

- He faxes *60 Minutes* producer Lowell Bergman, telling him he can't talk to him, and then faxes him again. (Motivator: politeness and curiosity)
- He goes to meet Bergman and asks about the consulting job. (Motivator: money for his family)
- In his nervousness, he lets slip that the report Bergman wants him to analyze is "just a drop in the bucket" and tells Bergman that he knows more than he's allowed to tell because of a confidentiality agreement he signed as head of research and development for a major tobacco company. (Motivator: honesty, integrity, and sincerity)
- He is threatened by his former boss but refuses to sign an extended confidentiality agreement. (Motivator: courage, sense of fair play, and righteous indignation)
- He moves into a smaller house and takes a job as a high school teacher to keep a low profile. (Motivator: care for family and caution)

153

꩜ He talks to Bergman again after Bergman tells him he didn't sell Wigand out to his former employer. (Motivator: contrition at having been too quick to blame)

꩜ Because he finds a footprint in his yard and thinks he's being watched, Wigand goes to meet Bergman to tell him what he knows. (Motivator: righteous indignation, courage)

꩜ We discover why Wigand lost his job: He refused to go along with the company's decision to leave a known carcinogen in its cigarettes because that chemical enhanced the nicotine delivery system that would continue to addict consumers. We realize what a really good guy he has been all along. (Motivator: integrity, commitment to right)

꩜ He receives death threats and struggles with the choice to go public on *60 Minutes*. (Motivator: worry for his family and their safety)

꩜ He agrees to appear on *60 Minutes*. (Motivator: sees importance of truth to public health)

꩜ He doesn't tell his wife what he plans to do and when she finds out and objects, he still does the interview because he knows they won't air it until they are sure it's legal to do so. (Motivator: misplaced loyalty to his word to network)

꩜ He agrees to give a deposition in a legal battle in Mississippi because it might get him out of his confidentiality agreement and allow the interview to be aired. (Motivator: worry about legal repercussions and the effects they may have on his family)

꩜ Even though his wife tells him " I can't do this anymore," he still refuses to discuss it until he gets back from testifying. He seems to believe that testifying is more important than his marriage. (Motivator: desire to honor the promise he made to testify)

꩜ He testifies even though he is told he might have to go to jail when he leaves Mississippi and re-enters Kentucky. (Motivator: belief in his cause)

꩜ He continues to answer questions even though he is warned by a Kentucky lawyer not to. (Motivator: belief in his cause)

꩜ When he gets home, he discovers that his wife and children are gone and yet he still wants his interview to air. (Motivator: wants to show his family what he did)

3. Write down the flaws your good character will have to overcome to solve the central problem of the film or to deliver its essential message and to achieve a satisfactory character arc.

Example from *The Insider*: Jeffrey Wigand's flaws:

- drinks too much
- quick to blame and quick to anger
- withdrawn and self-absorbed
- doesn't communicate with or confide in his wife
- altruistic to a fault—puts his family in jeopardy
  doesn't fight to save his marriage

These flaws create complications for Wigand. Because he drinks too much and is quick to blame and anger, he has a past that gives others cause to discredit him. His anger also motivates some of his decision to testify even at great cost to his family. His self-absorption adds to the breakup of his marriage. In fact, his altruism leads him into difficult situations where he must choose between what he thinks is his obligation to public health and his obligation to his family. All of these character flaws create an underlying tension and subtext to the decisions that fuel the plot. Wigand's character flaws also contribute to the difficulties faced by the film's other main character, Lowell Bergman. Bergman is forced to react to Wigand in ways that reinforce and support his choices. This makes for dynamic interaction between the main characters.

CREATING BAD GUYS

1. Make a list of your villain's horrible qualities.
2. List your villain's good qualities—those things that make him hesitate and will make him "lose" to the hero.
3. Decide what motivates the Bad Guy. Plot his bad choices based on malevolence or evil. These can serve to counter the "good" choices of your hero. Notice that they may alternate, that one may trigger the other, or they may exist separately.

Bad Guys' choices are usually based on their proclivities and desires and not on ideals or social pressures. For the most part, movie villains are victims of their own perversities. They usually act on compulsion and as a result of sensual desire and greed. Their mental perversity may be interesting at first but soon grows tedious because it is usually one dimensional. The more interesting you can make the choices, the more multidimensional your villain will be.

In *A Simple Plan*, Hank isn't really the villain of the piece. Greed is the real villain but Hank's reaction to and his *relationship* with greed is what turns Hank villainous. As an example, let's plot Hank's choices and what motivated them.

Sample Bad Guy Choice Chart: *A Simple Plan*

*The point of no return:
**first really evil choice.

Hank's choices:

- Lets Jacob's friend Lou come along to cemetery. Doesn't take him home first. (Motivator: laziness)

- Follows the guys into the snow to look for fox. (Motivator: wanting to be one of the guys)

- Goes into plane. (Motivator: curiosity and wanting to take charge)

- Agrees to keep money until spring and if no one claims it, they'll divide it up. (Motivator: wants control and greed)

- Covers up finding plane to cop. (Motivator: wants to keep money, fear)

- Tells his wife and "convinces" her they should keep money. (Motivator: greed)

- Goes back to scene with some of the money. (Motivator: fear of getting caught)

- Takes Jacob as a lookout. (Motivator: fear of getting caught)

- * Convinces Jacob to help him cover up "killing" of farmer to make it look like an accident. (Motivator: wants to get away with deed, no remorse, self-interest)

- **Kills farmer. (Motivator: to "protect" Jacob, to protect himself)

- Tells Jacob that he actually killed the farmer. (Motivator: to coerce Jacob to remain silent to protect himself)

- Tells wife about farmer and lets her convince him to keep it secret. (Motivator: self-interest and fear of getting caught)

- After he learns where the money came from, lets his wife convince him it's still not stealing. (Motivator: greed)
- Lets wife convince him to tape-record Lou confessing to farmer's death as protection against blackmail. (Motivator: self-interest)
- Convinces Jacob to betray his best friend Lou and help in taping. (Motivator: self-interest and greed)
- Tries to convince Lou's wife to "cover up" Lou's shooting. (Motivator: self-interest)
- Shoots Lou's wife. (Motivator: fear)
- Covers up shootings of Lou and wife by getting Jacob to lie. (Motivator: ruthless self-interest)
- Lies to cops about shooting. (Motivator: fear of getting caught)
- Doesn't heed wife's warning and goes with fake FBI guy. (Motivator: pride; thinks he can handle it)
- Takes gun from cop's cabinet and takes bullets before he goes out. (Motivator: caution and ruthlessness)
- Tries to warn cop. (Motivator: vestige of old kindness)
- Kills fake FBI guy. (Motivator: fear of getting caught, self-protection)
- Shoots Jacob. (Motivator: self-protection)
- Lies to cops. (Motivator: fear of getting caught)
- Burns money. (Motivator: self-interest, fear of getting caught)

In *A Simple Plan*, Jacob is really Hank's conscience. As soon as Hank commits his first evil act (killing the farmer), Jacob tries to convince Hank to do the right thing. He offers to go to the cops to report the farmer's murder but agrees not to because he wants to protect Hank; he doesn't want to trick his friend Lou into confessing to the farmer's murder but gives into his desire to get the family farm back; he offers to kill himself to get Hank to confess to the shootings. When Hank kills Jacob, he is, in effect, killing his conscience.

Certainly at that point and demonstrably throughout the film, Hank has no real remorse for anything he's done. He doesn't demonstrate remorse about any of the other killings in the film and neither does his wife. The woman who started off as a good person (wanting at first to give the money back) ends up being just as evil as her husband. She's got choices to make, too, and it's possible to create a Choice Chart for her. Sarah's choice chart is shorter than Hank's, but just as meaningful to her character.

Sarah's Choice Chart:

๛ Lets Hank convince her to keep the money.

๛ Convinces Hank to go back to the plane to leave some of the
   money.

*๛ Doesn't tell Hank he's evil after killing the farmer and hides that
   he did.

๛ Convinces Hank to tape-record Lou and get Jacob to help.

๛ After she finds out where the money came from, she convinces
   Hank to keep going with the scheme even when he wants to quit.

๛ Keeps silent about the whole thing.

   Sarah's prime motivation in every case: greed.

   Your Choice Charts will demonstrate the action flow of your film
and can be closely tied to a plot outline. Notice that Hank, in *A Simple
Plan*, by virtue of his choices, is actually a kind of villanero, but he is
not glorified. Instead, he is an instructive example of the corrupting
power of evil on good characters.
   If you consider only Hank's good qualities without knowing any of
the choices he makes in the film, you might consider him a Good Guy.
And yet, these good qualities might be considered "flaws" in the sense
that to commit to evil, Hank must "overcome," ignore, or make use of
these qualities to help him commit evil deeds. To demonstrate, I'll list
Hank's qualities and how they "helped" him become evil.

Example of Bad Guy "Flaws"
(In Hank's case some are actually "good" qualities.)

๛ Hank is honest and able to keep a secret.

๛ He's able to keep quiet about the money.

๛ He values freedom above anything.

๛ Because Hank doesn't want to go to jail, he continues to commit
   evil deeds to cover up his crimes.

๛ He is able to make a commitment to family.

๛ That commitment makes him want to succeed at any cost.

〜 He works hard and believes that work will help him get what he wants.

〜 He works very hard to make his scheme work and to stay out of jail.

〜 He is loyal, friendly, loving, intelligent, educated, trusting, and straightforward.

〜 He trusts his wife and his brother more than he should.

If you were to write a biography of Hank based on his character qualities, it probably wouldn't foreshadow his choices to do evil deeds. That's why it's important to also write *An Ethical Biography* that might give clues as to why even characters who are "basically good" might choose to do bad things.

ETHICAL BIOGRAPHIES

We've already talked about writing character biographies that might include facts about hobbies, relationships, and life events. We also need to create ethical character biographies that show what our characters' belief systems, moral precepts, and ethical concerns are, and how they have had direct bearing on their lives.

In Hank's case, for example, his ethical biography might have included examples of his selfishness stemming from his belief that he's better than his brother and more deserving of favors. In fact, morally and ethically, Hank probably believes that smarter and more educated people rule. And because of this belief, he probably values some people over others and believes that he has a responsibility to be friendly and law-abiding because he needs to set an example.

His ethical biography may also show that because Hank is so egotistical, he doesn't feel much empathy for other people, and believes it's okay to lie to protect himself and his family. Because Hank thinks so much of himself, he also believes that those who love him (like his wife) want what's best for him and he trusts them even though they may not deserve his trust. This kind of close examination of Hank's ethics might give some hint of the kinds of decisions he would make when under pressure.

I suspect that Scott B. Smith's biography of Hank would have been as colorful and evocatively descriptive as was the screenplay based on his novel. Ethical character biographies can be as creative and exciting as you want them to be, and should be written in the same style as basic character biographies, with an emphasis on image and events that might help you as you write your screenplay.

Begin your ethical biography in an interesting way. You don't have to start at your character's christening. You can instead start at an event that was a turning point for a moral or ethical stance taken by your character. Here's an example from an ethical biography of a character I created:

Ethical Biography of Good Guy Margaret

Tella and Margaret dreamed about peaches as they ignored history. It was hot in the classroom; there were beads of perspiration on their arms and elbows, and the tops of their scarred wooden desks were damp where they had leaned. They looked out the window and across to the fruit store, where they saw peaches—soft and fuzzy and plump peaches glowing in their bushel baskets. Their mouths watered as they waited for the dismissal bell to ring.

Minutes after it did, they were standing in front of the fruit store. Suddenly, Tella grabbed a peach, put it to her mouth, and bit into it. As the juice ran down her chin, she dashed away, eating as she ran. Margaret watched her go and then, inspired, grabbed a peach of her own, but before she could bite into it, she caught a glimpse of Mr. Marconi.

He was in the back of the store struggling with a heavy packing crate and as she watched him lift the crate onto the counter she suddenly saw how worn his hands looked and that his faded cotton shirt was stained with sweat. Carefully, Margaret replaced the peach. At that moment, she understood that stealing was a personal crime.

The description of this simple little incident makes it clear that Margaret will not steal later in life, but it also shows how sentimental she is, how observant, how tender-hearted and empathetic. And besides all that, when I write the screenplay, I can use the peach scene as backstory to a bit of business. During a conversation Margaret has with a friend in a supermarket, I could write that Margaret sees her friend pick up a fruit and put it to her lips. Without saying a word, Margaret takes the fruit out of her friend's hand and puts it back as she continues the conversation. It's a subtle little action, but it works to paint a picture of the character and to build up an ethical profile.

Exercise

Write an ethical biography (including several values) of about two pages, for the Good Guy characters in the screenplay you are writing.

You might even do this exercise before you outline or plot your story so that you can get some ideas for additional future actions or scene embellishments.

Write ethical biographies for your villains. Here's an example from one of mine.

Ethical Biography of Bad Guy Brian

The first time Brian tried to steal something, he was 9 and hanging out with his friends in Lou's cigar store. Lou was busy yelling at the other kids to get away from the girlie magazines so Brian was able to reach down to the bottom shelf of the candy counter and slip a Mars bar into his sock. He slipped out of the store alone and ran all the way home. That night, as Brian lay under the covers chomping on his chocolate, he thought how good it felt to steal something and get away scot-free. It was a feeling he'd try to recreate over and over in his adult life and one he remembered every time he tasted Mars bars.

Notice how the Brian character was different from the Margaret character. Brian was preoccupied with the act of theft. He wasn't particularly focused on the Mars bar. His thrill was in getting away with stealing. As a character, he wasn't sensitive. Instead, he was egotistical, self-preoccupied, and shrewd. This slipperiness can be used to great advantage in the screenplay. And so can all the possible business that can be derived from the candy scenario. In the screenplay, I can have the villain chewing on Mars bars after every heist; I can have him diabetic and longing to eat Mars bars; I can have him make his henchmen eat Mars bars; I can have him buying up candy stores and using them as crime fronts or money-laundering operations. There are lots of possibilities.

From these little examples you can see that the key is to create an ethical biography filled with scenarios that you enjoy. You should let your imagination go wild when you write these ethical biographies. Don't limit yourself and don't let the confines of your screenplay and story limit you. You can go beyond the box here and then perhaps, magically, what you write will expand the box and make it even more compelling.

Now that you know what choices your characters must make, what motivates those choices and what histories (ethical and actual) your characters have, you're ready to consider what actions they will take to demonstrate all of that information.

ANGELIC ACTS, DASTARDLY DEEDS

t's obvious to all of us that good characters perform "good deeds" and that bad characters do "bad things." It's simple enough. That is if you can tell good from bad. Ultimately, the thing that tells us the difference between the two is our conscience and the context in which deeds are performed. A character who is "neutral" will be passive and nonreactive but eventually will be forced to choose sides because even neutrality in a film causes something to happen. And what happens will determine if the neutrality was a good or bad "act." Again, context becomes important.

We're all familiar with examples of "neutrality" in news events. The most famous of these is, of course, the Kitty Genovese case in New York City. In that case, many people who heard Kitty cry for help as she was being killed in the street did absolutely nothing. Their "neutrality" resulted in a woman's death. In that case doing nothing could be looked on as "doing something"—contributing to a murder.

A movie example of "neutrality" or of a character doing nothing is *Leaving Las Vegas* (1994, written by Mike Figgis) in which Nicolas Cage plays Ben Sanderson, a drunk who does nothing to save himself from dying. His inaction (really negative action) creates misery for his girlfriend Sera (Elisabeth Shue) who wants to "save" him. In fact, his "neutrality"—not caring if he lives or dies—is self-destructive and excessively cruel to another person. (His girlfriend chooses to stay with him no matter what—an example of the Love Model.) In the context of the film, his neutrality seems almost immoral.

Context is really a film's environment . For example, what might be a good deed in one context, might be a bad deed in another. Let's take as an example the Three Gorges Dam Project flooding the Yangtze River Valley in China. To see that as a good deed, you'd want to think of all the electricity that water power would provide for the impoverished peasants of China. Villages that have never seen electric lights, that have no refrigerators, or any other modern conveniences, now could have the power they need to transform people's lives. Catastrophic flooding will be prevented, and China's economy will be improved because big ships would be able to navigate the river. To see the Dam Project as a bad deed, you'd have to consider the displacement of over a million people from homes they've occupied for generations, the destruction of some of China's best farmland, and the tragic loss of precious archaeological and cultural sites and dramatic natural scenery. Two different points of view, two different "act interpretations."

Are there acts that are always bad no matter what the context? To answer that we need to go back to our value systems. Do you believe that there are things so evil (or so good) they cannot be misinterpreted no matter the context in which they are performed?

What you believe will, almost in spite of yourself, come through in the way in which you write your movie. If, for example, you believe that evil is absolute and is clearly obvious in even the most blissfully good context, then your depiction of evil will come through as a strong entity in your film almost in spite of yourself. The same is true for the good acts you create in negative contexts.

The key is understanding the acts you have characters perform, what motivates these acts, and what effects these acts will have on your story. As a writer, you can't ever be wishy-washy about determining what your characters do. You've got to be clear about whether their actions are "good' or "bad" if you're going to create a successful and ethical film. And surprisingly, if you are clear about the nature of the acts your characters perform, you will be able to create film environments and characterizations that are also clear.

And as an added bonus, you will discover that when you have a bad character perform "bad" acts, they will be so obviously bad in their motivation and execution with regard to context, that you won't have to overkill the "badness" of them. For example, in a film where there is a lot of violence by both good and bad characters, often bad acts have to be especially graphic to indicate the depth of their evil. But in a film with less violence, bad acts stand out dramatically.

Any bad act by a character will indicate that character's ethical and moral belief systems. I call these acts *bad markers*. They can be graphic acts like battery or murder, or they can be toned down considerably (blatant lying, adultery, insensitivity, simple character quirks) if the action in the rest of the film is not violent and if the context of the film has clear definitions of good and evil.

Movie Example #1

Mission Impossible II (1999)
Written by: Robert Towne
Directed by: John Woo

Ethan Hunt (played by Tom Cruise) is an outrageously Good Guy who works as a spy. We are told how good he is right away, when he hangs, Christlike, from a ledge of a mountain he's trying to climb. Arms stretched out, feet together, and facing the camera, Hunt's image tells us subliminally that Hunt is someone who will save humankind. And his mission is to do just that by getting a deadly virus away from the Bad Guy, Sean, who wants to use it to rule the world.

We see how really bad the Bad Guy is when we're shown his "bad markers"; disguised as Hunt, he beats up a sick old man, and master-minds the takeover and subsequent crash of a passenger plane, killing all on board. By these "bad markers," the audience knows Sean's plenty bad and so does Nya, his ex-girlfriend, and the person Hunt needs to get into Sean's compound and find out what he's up to.

But once Nya's inside Sean's compound, the audience is shown a stronger "bad marker" to indicate how much danger she's in. Sean's so bad that he uses a cigar trimmer to cut off the tip of his loyal hench-man's finger when that guy disagrees with him. Even though Sean is shown being outrageously bad in the first act, the screenwriter be-lieves those "bad markers" are not conclusive enough to demonstrate what he might do to the girl if he finds out she is betraying him, and so the screenwriter goes for the gruesome stuff.

Movie Example #2

Strangers on a Train (1951)
Written by: Raymond Chandler, Czenzi Ormande,
and (uncredited) Ben Hecht
Directed by: Alfred Hitchcock

Tennis player Guy Haines (Farley Granger) meets Bruno (Robert Walker), a stranger on a train. Bruno is weird and more than a little nuts and offers to kill Guy's millstone of a wife if Guy will kill Bruno's hated father. Guy shrugs off Bruno's "plan" as the ramblings of a harm-less nut case. But Bruno—a twisted psychopath–actually goes through with his end of the deal and expects Guy to do the same.

Right from the beginning, there are "bad markers" to hint to the audience that Bruno might be bad and certainly mad. He's a little too friendly and familiar. He pries too much into Guy's personal life and he comes up with the "crisscross" murder idea. Every time we see Bruno, we're given more of these "bad markers": he displays an inordinate affection for his mother and she for him; he reacts inappropriately to his mother's horrible painting; his parents argue about committing him to an asylum; he stalks Guy. And right before he commits murder, he punctures a child's balloon with his burning cigarette. That rather obvious "bad marker" gives us a glimpse into how vindictive, mean, cold, and compulsive he can be. All Bruno's "bad markers" culminate with his ultimate bad act: the murder of Guy's wife, which we watch him commit in gruesome real time through the victim's broken glasses.

Just as you create "bad markers" to define evil for your audiences, you can also create *good markers*. Good markers are good acts performed by characters that define for the audience what you and your good characters believe about the nature of goodness. Good markers can be as simple as innocuous little character quirks: politeness, chivalry, sympathy, humor.

For example, in the first few minutes of *A Simple Plan*, lots of "good markers" are planted to show Hank is a good guy:

- he fills in for a friend who's late for work and is gracious about it;
- he's friendly to kids and dogs on his way home;
- he's jolly and friendly with the local cop;
- he is loving toward his pregnant wife and goes to the cemetery to place flowers on his parents' grave.

These little actions aren't really played up in the movie but are very essential pieces of character information.

What I'm suggesting is making movies more subtle and "deeper" by crafting controlled actions that are multileveled and suggest layers of character rather than blatant, simplistic actions (like those in some over-the-top action–adventure films) that scream so loudly that audiences are made insensitive and deaf to tone and innuendo. If writers take greater care to determine how characters can act elegantly but with definite purpose and results, stemming from their motivations and character traits, then the films they write will themselves be more elegant and, because they truly reflect the writer's intention and talent, they will be much more ethical.

One of my students wanted to show that his bad guy (a drug dealer) was really bad by sending the hero his mutilated pet cat. And this was supposed to be a *comedy*!!! Because I refuse to work on scripts that involve cruelty to animals, I told the student that if he wanted me to continue giving him feedback on his script, he should find another way of demonstrating the drug dealer's "badness." I suggested that the screenplay might be funnier and more subtle if the bad guy sent the hero a box containing only a cat collar and a tiny syringe. This wouldn't offend animal lovers and would still get the message across while it tied in with the "dark" drug dealer theme. There's no need to go to the most extreme and graphic images possible in order to accomplish what you want.

Writers do that because it's easier to be obvious than it is to be subtle. Our minds are lazy and most of us have pampered them by taking their first suggestions. We're too easily wowed by our own superficial brilliance. We've got to be tougher with ourselves and more demanding. If we push ourselves even a little, we'll come up with more creative, original, and interesting ways of saying things. And once our minds realize that we won't let them get away with junk, they'll start producing quality.

That holds true for ethics as well. If we're satisfied with making wimpy statements, our scripts will be wimpy. We've got to work to create dynamic stories that will demonstrate good and evil in original ways and that will describe evil without dwelling on it.

To describe evil without giving in to the urge to wallow in it is a remarkable exercise in control and sustained self-restraint. It's not an easy thing to do because, whereas it's necessary to show evil in order to reveal it for what it is, it's important not to spend too much time describing it. When screenwriters do that, evil's powerful imaging usually overshadows any positive message in a film. Of course, if your film is an exposé of evil, then you're going to have to describe that evil. But keep in mind that once you describe evil, instead of dwelling on it, you should quickly shift to a demonstration of the effects of evil—effects that you can use to bring home just how horrible it is.

One of my students wanted to show the inanity of war by writing a script in which medieval forces slaughter their way through a hostile European landscape. Over and over again, he had guys hacking off limbs, skewering bodies, raping, and pillaging. He created so much blood and gore that the screenplay became a crazy quilt of senseless brutality. (Imagine the gory first scenes in *Saving Private Ryan* going on for a whole movie!) Audiences watching a film like that would react first with shock, then horror, then revulsion, and finally would emotionally distance themselves from images too disturbing and relentless to engage them.

My student's film would not have delivered the message he intended it to. In fact, because of all the violence, the movie would have ended up looking like an advertisement for broadswords and tourniquets and nobody would have wanted to see that. I suggested that the student re-write the script by shifting focus to the effects of the battle on a variety of people: a professional soldier, a family in one of the towns, a woman following the army. When the student did that, as the "war" progressed, each person was gradually changed into despising what war did. In this way, with each battle, the film became more interesting because the audience could observe how war changed the lives of people they'd come to see as vital and "real."

When we write movies in which lots of violence or, for that matter, strong physical action occurs, we run the risk of succumbing to the temptation to ignore or sublimate character to action. It's important to remember that character is always more important than action because character determines action. And when evil actions are exaggerated or overblown and disregard character, then a story loses its vitality and its ethical center.

If we concentrate on our characters and "explain" them through action, we'll find that we can be more adept and creative at solving problems without floundering in violence, evil, and gore that may excite audiences but seldom enlighten them. And if we concentrate on our characters, and want to be socially responsible, we can show how they might suffer when they choose to do evil instead of good and we can still show evil but we will not be distracted into concentrating on it.

CRIME AND PUNISHMENT

t's an old saying that "no good deed goes unpunished." That means that ultimately, the hero who does good has to be confronted by the villain who wants to put an end to the hero's goodness. The "punishment" for doing good deeds is the struggle the hero must make to do them.

And what about bad deeds? Are they, and the villains who perform them, being punished as they should be? Not in most movies. Villains usually get away with their dastardly deeds or, if they meet spectacular ends (falling from great heights, getting blown up, eaten by sharks, or cut up in little pieces), those ends don't seem particularly memorable in the face of the prolonged evil we have to watch before good triumphs.

Probably that's because today's screen villains don't seem to suffer for their deeds in this world or the next. Screen deaths (even the gruesome ones suffered by Bad Guys), don't take that long, and the threat of Hell and the horrors of eternal damnation don't hold the same terror for sophisticated audiences as they did for medieval ones.

Consider Dante's *Divine Comedy*. Dante (1265–1321) spent lots of time in *The Inferno* section of *The Divine Comedy*, describing the hideous tortures inflicted on the damned. For example, he described in terrible graphic detail horrible wounds suffered by sword blows meted out as punishment on "instigators of scandal and schism." I'll omit most of the gruesome descriptions coming before the lines I've quoted about the actual punishment.

> *In front of me, Ali goes weeping,*
> *His face split open from his chin to his forelock.*
>
> *And all the others you see in this place*
> *Were instigators of scandal and of schism,*
> *When they were alive, and so they are split here.*

There is a devil behind here who hacks at us
So cruelly, with cuts of his sword,
And hacks again, everyone of our kind,

Every time we come round this road again;
Because the wounds close themselves up each time
Before anyone gets back to where he stands.[63]

Now that's something you don't want to see even in animation. The logic behind Dante's writing was that if he described the horror of the hell to which evildoers were subject (notice how the punishments fit the crimes), they would stop doing evil. That wouldn't work today. In our own times, we have debunked religion and the horrors of Hell so much that they are no longer deterrents. Capital punishment is not a deterrent. In fact, we seem more titillated than terrified by descriptions of the physical suffering of Bad Guys.

Perhaps then, we might finally move off the graphic descriptions of physical sufferings inflicted on evildoers into a more refined (and perhaps effective) description of internal suffering that evil deeds can cause. That works quite well. Take *The Sopranos*, for example.

During one episode of that HBO show, audiences see Tony Soprano suffer painful embarrassment when he is arrested in front of his daughter and her friends. Ordinary people can relate to Tony's painful emotions and gain some understanding of the emotional suffering a mobster lifestyle might inflict on mobsters. Of course, such examples lend themselves to comedy but there are dramatic ways of demonstrating how evildoers suffer and the agonies they endure without showing graphic violence.

Shakespeare did that in *Othello*. Othello listens to Iago and lets jealousy get the best of him. He murders Desdemona, his own true love, and then suffers horribly for it. We've already used the example of Salieri in *Amadeus* in our Good Guys section. Salieri effectively demonstrates the evil of egotism and jealousy by suffering greatly for being both egotistical and jealous. Here are some memorable examples from three great old movies.

❧ *Double Indemnity* (1944) Written and directed by Billy Wilder. Walter Neff (Fred MacMurray) experiences the agony of betrayal and guilt after he helps Phyllis Dietrichson (Barbara Stanwyck) kill her husband.

---

[63] Alighieri, Dante. (Sisson, C. H. Trans.). *The Divine Comedy*. Oxford University Press, New York, 1993, p. 164 (Inferno xxviii).

🙠 *The Postman Always Rings Twice* (1949) Directed by Ray Garnett and written by Harry Ruskin. Poor sap Frank (played by John Garfield) kills Cora's (Lana Turner) husband Nick (Cecil Kellaway) and suffers guilt and anguish.

🙠 *Rope* (1948) Directed by Alfred Hitchcock, written by Hume Cronyn and Arthur Laurents. Philip (Farley Granger) experiences unspeakable agony, fear, and guilt after he and his buddy kill a friend just for thrills and hold a dinner party with the corpse hidden in the room.

Suspense movies like those create their excitement by showing guilty characters sweating it out over being caught. But what if the movie you're writing demands some kind of active physical retribution? The key to that is creating the horror in the mind of audiences without actually showing horrible things.

In *The Boys From Brazil* (1978, written by Heywood Gould), we don't actually have to see Dr. Josef Mengele (Gregory Peck) get torn apart by vicious Dobermans. The grimacing face of Nazi hunter Ezra Lieberman (Laurence Olivier) is enough to let us know justice was served. The secret is to imply the punishment so well that it impresses itself on viewers' imaginations and satisfies their (and the characters') need for revenge.

No matter how they are punished, when villains lose, it's the resolution of the film's ultimate conflict—the struggle between good and evil. Too often, this conflict between good and evil is solved by violence, brutality, or force. In creating ethical films, it is essential that a broader scope be given to the ways in which conflict can be resolved.

Rollo May, in *The Courage to Create*, wrote that "conflict presupposes limits and the struggle with limits is actually the source of creative productions." He went on to explain that "creativity arises out of the tension between spontaneity and limitation"[64] and we can use that same principle when it comes to choosing the kinds of limiting conflicts we create for our characters. If conflicts are limitations, then we can use our creativity to come up with different ways in which our characters can move forward beyond these limitations.

Our characters don't always need to beat their heads against walls any more than we do during the writing process. Just as we've got to find original ways of solving our creative dilemmas in the writing process, so, too, do we have to find original ways to get our characters to solve their conflicts and overcome their limitations.

[64]May, Rollo. *The Courage To Create.* Bantam Books, New York, 1978, p. 137.

The kind of conflict you will demonstrate in your screenplay depends largely on the type of film you are writing. Stories with internal or relationship conflicts can usually be resolved without graphic violence. Action–Adventure films most often require some kind of visual violence. Comedy is in a particularly exclusive category. It can include violence and violent acts as satire (e.g., *Monty Python and the Holy Grail*, 1975) and usually a question of taste determines how far these films can go in showing graphic acts.

SPECIAL CIRCUMSTANCES

FUNNY OR NOT, HERE I COME!

People take comedy too seriously. I am always surprised when someone is offended by a comedy and that's because it's the very nature of comedy to be irreverent, and sometimes even annoying, getting in your face and under your skin. Sometimes, comedy is even meant to offend. But when comedy doesn't work, it's deadly. That's why it's so difficult to write. It's hard to create concepts that take people by surprise, engage their enjoyment of the ridiculous and absurd, and walk the edge of impropriety just enough to make them laugh but not enough to turn them off.

I'll put up with a lot in comedy because I believe funny bones are very personal. What makes one person laugh might have no effect on someone else. Personally, for example, I believe some subjects should be off limits. As far as I'm concerned, child molestation, animal abuse, racism, and misogyny are not funny. Sometimes ethnic, religious, and physical stereotyping are used in movies as gags and joke hooks. That's when things get precarious. I don't believe overtly derogatory ethnic, religious, and physical stereotyping is funny, but then there are those times when a screenwriter, as part of a particular ethnic group, does use it to good advantage. Remember the hilarious scene in *Annie Hall* (written by Woody Allen and Marshall Brickman) when a Jewish Alvy Singer (Woody Allen), having dinner with Annie's (Diane Keaton) WASP family, suddenly takes on the persona (complete with earlocks and outfit) of an Hasid sitting down to eat a pork roast. A stereotype, but nevertheless, funny!

When you're writing comedy, you've got to be bold, brave, and outrageous, and sometimes that means taking chances that might not work. Fortunately, the moment you take an irreverent and light tone when you tell a story, you commit yourself to the idea that you are being silly and what you say is all in good fun.

172

Look at Jerry Zucker's *Rat Race* (2001, written by Andy Breckman). In that film, a desperate guy (Jon Lovitz), in a race for big bucks, is nagged by his little girl into making a detour to visit a Barbie Museum. We all think (as the family does) that the museum will be filled with all things Barbie—dolls, accouterments, Ken.

It's absolutely hilarious when we find out that the museum is actually a Nazi shrine dedicated to the memory of Klaus Barbie, the man once called "The Butcher of Lyon." The family members (and they are Jewish) pretend to love Nazis as they are led through the museum by skinheads. They finally make their "escape" in what used to be Hitler's touring car. That gag sequence alone is worth the price of admission. Could it offend some people? Perhaps, but the sequence was intended to be genuinely funny and it worked.

Now we all know that the Holocaust wasn't funny, so people writing Nazi jokes have to know how far to go and must display good taste. They've got to make fun of Nazis without making fun of atrocities they committed. Charlie Chaplin took chances with *The Little Dictator* and was sensational. Mel Brooks took chances with the song "Springtime for Hitler" in *The Producers*. He was fabulous, too.

When you're writing that kind of comedy, your refined sensibilities and conscience come into play. You'll probably be able to sense when you're off. If not, try your humor out on people you respect and who will tell you the truth.

It's too bad that the Farrelly brothers didn't do that when they made *Shallow Hal* (2001, written by Sean Moynihan and the Farrellys). The movie's about Hal (Jack Black), a regular crass guy obsessed with beautiful women until he is hypnotized by Self-Help sadhu Tony Robbins (playing himself) into seeing the true "inner beauty" of women. His hypnotic trance helps him fall in love with Rosemary, a 300-pound woman he "sees" as the svelte Gwyneth Paltrow.

The Farrellys claim that the movie's message is that appearances shouldn't count when it comes to true love. They meant well and I really do believe they intended to say something important and "deep." Unfortunately, their good intentions were overshadowed by the insensitivity, mean-spirited humor, and poor taste of their movie.

Most of the jokes in *Shallow Hal* are at the expense of the calorically challenged and at women who look less than perfect. And it tries so hard to be true to its message and light-hearted at the same time that even the disabled are forced to make jokes about their own conditions! That's profoundly embarrassing and uncomfortable to watch.

In fact, there are so many mean jokes (unattractive women are called dogs and hyenas, fat women are called hippos and rhinos and are shown destroying the furniture they sit on) that it seems like the

Farrellys themselves are trying too hard to overcome what may be their own male proclivities for perfection in the appearance of women. Proof of that? Hal has to work pretty hard to overcome his disgust and aversion to corpulence.

If the Farrellys truly feel that inner beauty is more important than outer beauty, they wouldn't have been able to bring themselves to write this kind of destructive humor. But their prejudice is clear. In *Shallow Hal*, they equate ideal inner beauty with thinness—an equation that has been scientifically documented as contributing to eating disorders and much misery among girls and women.

Shallow Hal received flak from obesity acceptance groups, some of whom called for a boycott of the film, and yet the Farrellys aren't fazed. Peter Farrelly says "he and his brother always listen to criticism, but the controversy won't change the way they write future projects. 'You can't write out of a sense of fear,' he said. 'You write and listen to your own god and stay true to it. And that's what we did.'[65]

Peter Farrelly is right. When you write comedy, you have to first think seriously about your message, do lots of research, take stock of your own deepest thoughts about it, and then take your best shot. You have to take chances and you have to be bold, brave, and fearless.

Unfortunately, sometimes, even if they don't mean to, comics will poke unfunny fun at someone and something just like the Farrellys did. When they do that it's the responsibility of audiences to register disapproval by not laughing and by staying away from theaters. Consumers have more power than they realize and always determine if comedy works.

Sometimes, screenwriters who aren't sure just how nasty or appropriate their comedy is will claim that it's dark. And conversely, screenwriters often try to mitigate superviolent or potentially offensive material by giving it a funny spin.

For example, some of the conflict resolution in Quentin Tarantino's *Jackie Brown* was pretty violent but it was also funny. When Louis Gara (Robert De Niro) was being driven crazy by Ordell Robbie's (Samuel Jackson) girlfriend Melanie (Bridgett Fonda) in the department store parking lot, Louis takes out his gun and suddenly shoots her dead. That's pretty outrageous but also quite funny—I'm sure Tarantino intended it to be—and that humor ameliorates the extremity of the violence. Some people who write violent movies do that as a way of stepping back from the violence and making it more palatable but still, it's pretty hard to watch and may even

---

[65]Smith, Lynn. *Does "Hal" Send Mixed Signals?* The Los Angeles Times, November 17, 2001, p. F13.

be exploitative and unethical. Remember *A Clockwork Orange* (1971, written and directed by Stanley Kubrick) where people are beaten to death to happy music? Alex (Malcolm McDowell) even rapes a woman while crooning "Singing in the Rain." I didn't laugh during that graphic scene! In fact, it almost made me physically ill.

Contrast the violence in *A Clockwork Orange* with the violence in *Fargo*. The violence in *Fargo* is hard to watch, too, but the film is really funny and, in fact, the quintessential example of the kind of comedy that comes from situation played out through character. Just as in drama, if you pay attention to your characters in comedy and if they are innately funny and quirky, your screenplay will be funny, too, even if it's about bad deeds.

Movie Example

Fargo (1996)
Written by :Ethan and Joel Coen
Directed by: Joel Coen

In *Fargo*, Jerry Lundegaard (William H. Macy), a selfish and suppressed car salesman, is driven to do a dastardly deed by his intense desire to get out from under the thumb of his nasty rich father-in-law (also his boss). He hires two thugs to kidnap his wife in the hopes that his father-in-law will pay the ransom to rescue his daughter. Lundegaard expects to pay the thugs off, make sure his wife is safe, and end up with the lion's share of the ransom money, with which he can fund the financial scheme that will make him independent and rich.

Naturally things don't go as planned. One of the two thugs (Gaear Grimsrud, played by Peter Stormare) turns out to be a trigger-happy and cold-blooded psychopath. He ruthlessly kills three people on the way to the hideout with the kidnap victim. The other kidnapper, Carl Showalter (played brilliantly by Steve Buscemi), a slimy motormouth who tries to act mean, winds up killing the father-in-law and making off with all of the ransom.

And hot on their trail is Marge Gunderson (Frances McDormand), a very pregnant local cop who takes a practical no-nonsense approach to her manhunt. Marge's simplicity and deadpan sincerity is endearing, as is her relationship with her decoy-painting, unemployed house husband.

Fargo is a very bloody and violent film that has the potential of being horribly bleak. Instead, the Coens skillfully transform the movie into a really funny and powerful dark comedy. To do that, they create characters who are intrinsically amusing and a joy to watch. Every character in *Fargo* is quirky and outlandish. Not meaning to be funny, they are hilarious by virtue of their preoccupations, mannerisms, and speech patterns. The Coens make the most of the Minnesota environment and regional peculiarities of Minneapolis to punch up the comedy. Even the horribly evil Bad Guys are funny because they are so peculiar.

Particularly hilarious are Madge's interview with two teenage hookers and the exchange between a cop and a guy shoveling snow. In both these scenes, the cops are being given vital and helpful information about the Bad Guys but the exposition is beautifully and artfully buried in the fabric of homespun aphorisms and chatter.

The Coens end *Fargo* with a moral message ("There's more to life than money") delivered by Marge in a kindhearted motherly lecture to the surviving Bad Guy captive in the back of her police cruiser. Because it's the kind of thing that a character like Marge would do, the speech doesn't come off as preachy but instead clues us into the intrinsic values of the Brainerd community and of Marge herself. And so it's kind of sweet and, because it's falling on deaf psychopathic ears, it's also funny.

Screenwriters who think that writing "dark comedy" gives them the right to throw in as many gratuitous acts of violence as they like, fool no one. It takes great skill to write real "dark comedy." It takes an understanding of both comedy and drama and a delicate hand and sensibility. Unfortunately, not many people are capable of that. Quentin Tarantino and the Coen brothers may be good at it, but if you're wise, you'll stay away from "dark comedy" until you have the experience and skills to tackle it. Otherwise your screenplay could just be a series of evil acts portrayed too lightly, and that might give you the appearance of being a callous, cruel, insensitive, and unethical screenwriter.

CAPER FILMS: BARELY DARK

Caper films are about criminal acts. The goal of the caper is usually a spectacular and/or difficult theft. Villaneros in caper films have to pull off the impossible and have to break the law to do it. So are caper films unethical? Certainly not. And that's because they aren't *really about* stealing and the glories of stealing. They're *really about* the characters and personal relationships of the thieves. In fact, character and relationships are the things that make caper movies interesting. That, and the ingenious cleverness of a caper's plan—ingenious cleverness that adds to our knowledge of and interest in the characters.

Think of all the great caper movies you've seen and how much fun you had watching them. Classic gems like *To Catch a Thief* (1955), *Topkapi* (1964), *The Sting* (1973), and many more of that ilk, all take a light-hearted approach to criminal acts and do make heroes out of thieves (villaneros). *But* they do so with wit and style, concentrating on the characters of the thieves and their relationships with each other. Caper films are unethical and socially irresponsible *only* if they are mean spirited, humorless, and hideously violent.

A caper film that relies on humor downplays any message that stealing is good. In fact, most caper films do hold capture and punishment over the heads of the thieves and that adds to the suspense of the movie. And lately, much prominence has been given to an insurance coverage rationale for stealing. Insurance coverage is discussed in *Bandits*, where banks are being robbed, and even in *Ocean's Eleven*, where the goal of the caper is the robbery of a Vegas casino.

Movie Example #1

The Score (2001)
Written by: Kario Salem, Lem Dobbs, and Scott Marshall Smith
Directed by: Frank Oz

Robert De Niro plays Nick Wells, an aging jewel thief who owns a jazz club on the side and is having a torrid romance with a flight attendant (Diane, played by Angela Bassett). Anxious to deepen his relationship with her, and tired of pulling jobs, Nick decides to get out of the burglary business, but is talked into pulling one last caper by kingpin fence and long-time associate Max (Marlon Brando).

The caper's been devised by Jack (Ed Norton), a young thief wannabe. Jack gets work as a caretaker at the Montreal Customs House, where he pretends to be the feeble-minded Brian. Trusted by everyone because they think he's retarded, he's free to snoop. That's how he discovers a precious ancient French scepter captured in a botched smuggling attempt, stashed in a vault. The caper concentrates on pinching the scepter (it has a nefarious history and inconclusive ownership), before it gets sent back to the French government.

Jack/Brian has the inside lead—access to building schematics and keys—but he needs Max for money for burglar knickknacks, and Nick for his expertise in breaking into impossible places and opening impenetrable safes. Nick doesn't trust Jack from the beginning, but is lured into the scheme by the promise of big-time loot and by the fact that his friend Max will be killed unless he comes up with big money (his take from the scepter caper) to cover a debt to the mob.

Naturally the relationship between Nick and Diane becomes strained when he breaks his promise to her and does this one last job. His relationship with Max doesn't fare well either—especially as Nick's "partnership" with Jack becomes increasingly problematic.

The Score is really about relationships, loyalty, and respect. The movie's message is that young people (even thieves) need to respect older, more experienced people. (We talked about that earlier.) The caper is interesting because we get to see just how smart Nick is, and how he handles the dynamics of his relationships with Diane (the flight attendant), with Max, and with Jack/Brian. All these things, plus the "victimless" nature of the crime, makes The Score an example of the ethical caper film.

Movie Example #2

Bandits (2001)
Written by: Harley Peyton
Directed by: Barry Levinson

In *Bandits*, ex-cons (and villaneros) Joe (Bruce Willis) and Terry (Billy Bob Thornton) get into the banks they rob by sleeping over at the bank manager's house the night before a heist. These "Bad Guys" are pleasant and funny; bumbling buffoons who endear themselves to those they rob by doing bad things in a way so quirky, dumb, and cute they can't be called evil.

Joe even explains to one bank manager his sliding scale rationalization for robbery: that Joe and Terry aren't stealing the bank's money because that money is insured by the government and the government steals from all of them. At the end of *Bandits*, Joe and Terry get away with the loot and we enjoy watching them do that because they are so adorable and funny. We don't really want them to be punished for their crimes.

Bandits is in the true tradition of most caper films that (even though they are about unethical acts) are pretty enjoyable and "harmless" because the capers are interesting and don't seem to hurt anyone "good" and especially because the "bad guys" are engaging and not too "bad."

Movie Example #3

Heist (2001)
Written and directed by: David Mamet

Whereas *Bandits* fits the caper formula, *Heist* is so dark it straddles the line between caper and action–adventure/thriller genres. That's because, although *Heist* is still about characters and their relationships, those characters are smart, mean, hard, and ugly.

Veteran super-thief Joe Moor (Gene Hackman) and his much younger shill of a wife Fran (Rebecca Pidgeon) decide to hang up their burglar tools after one last complicated, mob-engineered job. Throughout the film, Joe's actions and those of his cronies seem bad because none of the characters are likeable. Even though they are quirky and even though they steal from rich people (with insurance), a big Swiss bank (with insurance), and really really Bad Guys (who cares if they have insurance??), they still seem evil. It doesn't look like anyone is having any fun.

At the end of *Heist*, Joe gets away with the loot but those of us who didn't like *Heist* aren't too happy that he does because his escape is preceded by a fierce and bloody shootout that leaves lots of people dead. This is a little too dark to make the caper enjoyable.

At the end of *Heist*, we're told that the price of getting the loot is the violent killing spree, the life of the most likeable Bad Guy Pinky, and the betrayal of Joe by his wife. That's a direct statement about how evil gets punished and it mitigates the potentially "unethical" nature of the film. We're led to believe that the hero isn't happy in spite of his final "win," but because everyone in the film is so ruthless and nasty I'm not too sure we believe Joe's success is a Pyrrhic victory.

I think that Mamet's statement about the cost of the caper in *Heist* is a little too on the nose. Whereas it's good to show that villains (or villaneros) are punished, it's also important for screenwriters to avoid being preachy or didactic when setting up and talking about punishment. When screenwriters are too preachy, they alienate audiences from their messages. If you want to make films about something important to you, it's important that you let your characters and story present your message for you in ways that will keep audiences involved in that story. That way, your message can hit home subtly—almost subliminally—by taking hold in an audience's subconscious mind through its emotional and intellectual involvement in your movie. It's always important to remember you're a screenwriter, not a minister.

PART V
KILLING THE MESSENGER

NO SERMONS

I t's always a sad thing when sincere screenwriters who want desperately to do good come off looking like pompous religious fanatics or ethical blowhards. That's particularly a problem for "religious" screenwriters who can't seem to curb their enthusiasm for Biblical maxims and make their characters quote scripture. Nothing turns audiences off more, particularly if they are unconverted to the screenwriter's point of view. And because it's the "unconverted" that most screenwriters concerned with ethics (and/or religion) are writing to influence, they defeat their own purpose.

Screenwriters who want to make movies specifically to evangelize should know that they are creating propaganda instead of entertainment. When plot conflicts or serious problems faced by characters are magically erased by prayer or sudden conversion, audiences will know that something is not quite right in movieland. They'll feel duped, hoodwinked, and manipulated instead of uplifted and inspired.

Movie Example #1

Left Behind (2000)
Written by: John Bishop and Joe Goodman
Directed by: Victor Sarin

Here's an example of how a story that might have been a good action–adventure film is taken seriously off-track by heavy-handed proselytizing. It's based on the mammoth best-selling series of Christian novels about the Rapture and resulting fight with the Antichrist.

In this story, the Antichrist is a guy who heads up the United Nations and is interested in creating a global economy and government divided into ten "provinces." And his game plan has apparently been foretold by Biblical prophecy.

Left Behind (a made-for-video movie complete with weak acting and even weaker writing) is a classic religious propaganda film. One of the main characters, pilot Rayford Steele (Brad Johnson), is having an affair with a flight attendant and neglecting his Christian wife and son. He comes home and finds his wife gone (as one of the "chosen true believers," she's been taken to heaven in the Rapture). He falls on the floor at the side of his wife's empty bed, bemoans his fate, and quickly locks eyes on her bedside Bible. Sobbing, he first hurls it against the mirror but then, almost immediately, he picks it up and begins to read it.

"Converted," he rushes to the church, falls on his knees and prays and then suddenly becomes a "new" person and clutches the Bible throughout the rest of the movie. Sanctimoniously calm and overtly holy, he goes about converting his daughter and, ultimately, Buck, the nosy reporter (Kirk Cameron) trying to save the world from the Antichrist. We're treated to the scene of Buck's conversion: holed up in a bathroom, he sinks to the floor and finally talks to God. After that less-than-moving episode, Buck joins the pilot and his daughter, giving the audience a big clue that their combined faith will help to overthrow the Bad Guy whose dark deeds threaten to play themselves out in many sequels.

All the dramatic kneeling and praying out loud in *Left Behind* is heavy-handed and obvious and turns off viewers who would prefer to be led more gently (and powerfully) through story and character development to the conclusion that faith in God and goodness gives people the strength to cope with and to overcome evil.

Movie Example #2

The Omega Code (1999)
Written by: Stephen Blin and Hollis Barton
Directed by: Victor Sarin

The *Omega Code* seems to rehash many of the plot points in *Left Behind* but I suppose that's because they are both based on the same Antichrist-will-take-over-the-world prophecy. A much bigger budget and the addition of some bigger "name" actors (Michael York and Catherine Oxenberg) try to add some excitement to the movie. And in fact, so does the premise: the discovery of a secret code in the pages of the Bible that foretells world events.

This *Omega Code* is discovered by a failed priest who's friends with billionaire industrialist and Antichrist figure Stone Alexander (played by Michael York). Alexander plots to use the code to control the world by taking over (once again!) the United Nations, creating a global economy and a global government. Along for the ride is gullible and power-hungry self-help guru Gillen Lane (Casper Van Dien).

Just like the pilot in *Left Behind*, this guy ignores his loving wife and child to go off to exciting work (a pattern for guys' pre-conversion?), but in this film, he unknowingly teams up with the Bad Guy who uses him to help motivate people to join the world globalization effort.

The *Omega Code* is a little more subtle than *Left Behind* in conveying its message but it still has mysterious preaching prophets and scripture quotations presented in voice-over and as part of groovy special effects to make them more palatable. It doesn't work. The film is still too preachy and plays off that tendency by showing us the ultimate "religious" moment. The disillusioned Gillen is imprisoned in a jail cell by the forces of Satan, but when he falls to his knees and asks Jesus to save him, the cell door magically opens and he walks out a free man.

In *The Omega Code* we're treated to more special effects, "historical" information and location shooting. We also see intrigue, Israeli commandos, war rooms, and Satan taking over the body of the mortally wounded Stone Alexander. But alas, in all this heavy-hitting action–adventure drama, we have trouble believing that the hero's conversion is enough to get him (or the world) out of trouble.

All these "religious" films that feature prominent kneeling and loud praying might take a hint from the scriptures they are so fond of quoting. As it says in Matthew 6:5:

> And when thou prayest, thou shalt not be as the hypocrites are, for they love to pray standing in the synagogues and in the corners of the streets, that they may be seen of men. Verily I say unto you, they have their reward. But thou, when thou prayest, enter into thy closet, and when thou hast shut thy door, pray to thy Father which is in secret. And thy Father which seeth in secret shall reward thee openly."[66]

In the case of "religious" movies, the reward of "closet" or subtextual prayer, instead of overt preaching and public genuflection, might be audience satisfaction.

Movie Example #3

The Prophecy (1995)
Written and directed by: Gregory Widen

---

[66]*The Holy Bible*, Thomas Nelson and Sons, New York, 1952, p. 760.

The *Prophecy* is an interesting example of a "religious" film disguised as a horror movie. I say disguised because it takes an approach that seems to reject goodness and glorify evil. In fact, it preoccupies itself with evil and portrays it splendidly because it's helped by the world-class acting of Christopher Walken as Gabriel (the Bad Angel jealous of humans) and Eric Stoltz as Good Angel Simon. Boffo actor Amanda Plummer even makes an appearance as a person used, abused, and bound for purgatory. And there's lots of money on the screen—special effects, great cinematography, interesting writing.

The theme of *The Prophecy* is the war in heaven between the angels as it plays out in front of Martin (Elias Koteas), a priest who has lost his faith and turns to police work. And the movie makes no bones about showing just how horrible disgruntled angels can be. Even the good angel Simon (played by Eric Stoltz) is pretty awful: He gouges out eyes and brutally kills or wounds attackers. Much of the usual sermonizing here is obscured by churchy dark music, tough guy talk, experimental image flashes (also seen in *The Omega Code* but used to better advantage here), and general horror-genre gruesomeness and gore.

We get to watch souls being sucked out of people, souls being battled over, and native American ceremonies used to exorcise demons. We even get to see Lucifer (played by Viggo Mortensen) showing up to make deals and tempt the Good Guys. And in spite of all the pointed attempts to be dramatic and chilling and nonpreachy, the film still quotes scripture (in the mouths of superb actors it manages to play) and sermonizes at the end with the main character declaring "I have my soul and I have my faith."

The images *The Prophecy* presents are so graphic, and the violence so predominant, that it falls firmly in the camp of films that religious people often decry and avoid. In spite of its proposed affirmation of faith, it winds up, albeit unintentionally, being preoccupied with evil, and evoking and even invoking it.

These movie examples show that it's not an easy thing to talk about religion, faith, prayer, and God in a film and make the film palatable. It's always better to have characters demonstrate their faith by compelling and motivated action instead of having them speak words or perform conspicuous acts of drippy devotion.

We've already seen that messages are important and endemic to films, so just how does a screenwriter convey a message (religious or not) he or she cares deeply about and keep it subtle? The trick in doing that is to first find out what your message is, arrange your scenes so that your story makes your point, and then forget about your message. Concentrate instead on story, character, and making your scenes work, and don't worry about hitting people over the head with what you want to say. If you have taken the time to know what you think, to create an interesting story that demonstrates that deep characters whose actions are motivated by their good and bad qualities, then your message will get through.

Here's my preachy message for this chapter:

The purpose of characters and their dialogue is not to deliver the screenwriter's message but to live their lives on screen. In the course of that living, "messages" will be delivered. These "messages" usually coincide with what the screenwriter actually believes, but that should not be obvious to the audience.

Take a lesson from all of the world's great religions. Every one of them uses stories to transmit truths. The Book of Matthew in the Christian Bible, for instance, tells us "All this Jesus said to the crowds in parables; indeed he said nothing to them without a parable."[67] The prophet Muhammad tells us "we have given mankind in this Koran all manner of parables, so that they may take heed."[68] You'll find similar references to parables and stories in the sacred texts of Jews, Buddhists, Sufis, and more. They all know that the best kind of "teaching" is done through stories and not dictums that jab people in the tummy.

That's something the Farrelly brothers can't resist doing in *Shallow Hal*. Even thought it calls itself a comedy, that movie is so preachy that it actually has Tony Robbins delivering the message almost right to camera. That preachiness and overt messaging drives yet another nail into the grim coffin of that dreadful comedic failure.

If you have to include action that is "messagey," keep it brief—particularly in comedy. At the end of *Rat Race*, for example, lots of audience members who otherwise loved the movie were turned off by the heavy-handed (albeit good) message delivered when the racers turned all their money over to a world food organization. That great message would have been better served if the scene had been much shorter. There was far too much time spent getting the racers to give up their money, getting the Bad Guy gamblers to cough up for the cause, and then moshing in joy at the whole notion of contributing to food banks.

Please!! Resist the urge to fling yourself into a self-congratulatory mosh when you write! Great messages crash and burn when zealous

[67]Ibid., p. 768.
[68]Dawood, N. J. (Trans.). *The Koran*, Penguin Books, London, 1997, p. 324.

filmmakers and screenwriters can't resist the temptation to celebrate their own commitment to goodness.

Kids' movies are particularly susceptible to preachiness. But at least the messages in those movies are usually couched in cute songs and light-hearted banter. The movie *Babe*, for example, is quite effective pro-vegetarian propaganda made palatable by the adorable fretting of barnyard beasts during potential killing sprees by hungry farmers. The scene where the duck watches through the farmhouse window as his "friend" Roseanne is served up for Christmas dinner delivers some really dark humor and the big message (Animals are people, too), both at the same time. I bet most people watching that scene at least think about never eating meat again! And they are entertained into doing that by heartwarming characters that make the movie (and its message) fun.

Lots of the old Disney movies famous for their preachiness are forgiven for sermonizing because they are so intensely cute. And they contain memorable songs to boot. Let's face it, gullible and impressionable kids may be fodder for big message making but these days those messages have got to be delivered in a really sophisticated way. Take the "friendship" and "loyalty" message in *Toy Story I*—heavy but adorably delivered by heartwarming characters. And the message in *Monsters, Inc.*—that laughter is more powerful than fear. Engaging and rollicking fun, again dished out by cuddley heartwarmers!

But take heed. Although big message making might work in movies ostensibly designed for kids if the delivery systems are cuddly and precious, I can't think of a single movie star cuddly or precious enough to make big messages palatable to adults. I know, I know. Some of you are probably shouting out the names of your favorites even as you read this, but in spite of your fond fetishes for pretty faces, I still don't think that audiences these days want to see their superstars sermonizing. In the old days, Charleton Heston might have been able to get away with those Moses lines, but these days it takes more than a funky costume and a skyward gaze to put the fear of God into audiences.

I know it's hard to resist the temptation to go for the big message punch when the opportunity presents itself to do so, but resist you must. And you've got to especially resist the temptation to have characters talk for you and say directly what you want the movie to say. All those dialogue opportunities may be lying there, just waiting for you to use them, but the more obvious they become, the more you have to practice self-restraint.

Exercise

Watch movies you think have a particular "religious" or "spiritual" message and take note of what you think worked in these films. Because religion is so personal, you can look at films that were meaningful to you. Here are some suggestions:

- *Dogma* (1999, Written and directed by Kevin Smith. A comedy featuring renegade angels.)
- *The Apostle* (1997, Written and directed by Robert Duvall. Duvall plays a fallen preacher.)
- *The Last Temptation of Christ* (1988, Written by Paul Schrader. Directed by Martin Scorsese. The life of Jesus based on Kazantzakis' wonderful [and controversial] novel.)
- *Resurrection* (1980, Written by Lewis John Carlino. Directed by Daniel Petrie. Ellen Burstyn plays a faith healer.)
- *Brother Sun, Sister Moon* (1972, Written by Suso Cecchi d'Amico and Kenneth Ross. Directed by Franco Zeffirelli. The life of St. Francis.)

WORDS OF WISDOM

Self-restraint is, in fact, the key to writing good dialogue. Dialogue is a really difficult thing to write. Some people have a gift for it but most just find it torturous going. That's because every character in a screenplay needs to talk in a way unique to that character. If we wind up having every character sound the same, it will seem as if dialogue is only a vehicle for the writer to talk to the audience. That's something screenwriters want to avoid making obvious because that approach takes the audience "out" of the movie. A great writer creates characters so real and compelling, who talk so engagingly, that audiences forget they're not real and totally give themselves over to the movie and, ultimately, its message.

Sometimes actors bring so much to the dialogue that its impossible to differentiate between a star and the character played by that star. You see that a lot in John Wayne movies. Wayne always plays Wayne, Eastwood is always Eastwood, and Willis is usually Willis. When writers write screenplays with stars in mind, they tailor the dialogue to sound like those stars sound. It's a rare star that can be chameleon enough to overcome the "curse" of his own larger-than-life persona and transform into different people. Laurence Olivier had that ability. So does David Souchet, Billy Bob Thornton, Meryl Streep, Jennifer Jason Leigh, and a long list of character actors we don't even know. The trick for writers is understanding each character and defining that character so completely that writing his or her particular dialogue becomes, if not an easy, then at least a comfortable thing to do.

But that isn't the only difficulty in writing dialogue. It's also important for the dialogue in a film to be subservient to the action. We need to see characters do a thing instead of talking about what they are going to do or what they have done. And we don't need to have things explained to us. Exposition is the death knell of a script. Sometimes, on very rare occasions, it's necessary, but when that is the case, dialogue should always be written as short as is possible to write and have it still make sense.

Casablanca (1942), written by Julius and Philip Epstein and Howard Koch (and, it's rumored, nine other writers) is the classic example of brilliant "expository" writing. There are plenty of places where the audience is given facts about a character, but these are delivered surrounded by wit, and great writing. Consider the following lines where we learn several things about the mysterious Rick Blaine (Humphrey Bogart): that he's brave, adventurous, and worked as a mercenary. We also learn that he has an altruistic streak that he tries to hide, and that information will be important at the end of the movie when the audience needs to believe he'll give up the girl for a greater cause.

> CAPTAIN LOUIS RENAULT
> In 1935, you ran guns to Ethiopia. In 1936
> you fought in Spain, on the Loyalist side.

> RICK BLAINE
> I got well paid for it on both occasions.

> CAPTAIN LOUIS RENAULT
> The winning side would have paid you much better.

More expository wit from *Casablanca*? When Major Strasser lists Rick's particulars and shady background, Rick's sardonic comment, "Are my eyes really brown?" softens the exposition and quickly becomes a trademark for his character. The audience delights in his rapier deadpan wit and that makes for not only an endearing character but for a screenplay with great and memorable lines. Who can forget these great exchanges:

> CAPTAIN LOUIS RENAULT
> What in heaven's name brought you to Casablanca?

> RICK BLAINE
> My health. I came to Casablanca for the waters.

> CAPTAIN LOUIS RENAULT
> The waters? What waters? We're in the desert.

>   RICK BLAINE
> I was misinformed.

And:

>   MAJOR STRASSER
> Are you one of those people who cannot imagine
> the Germans in their beloved Paris?

>   RICK BLAINE
> It's not particularly my beloved Paris.

>   HEINZ
> Can you imagine us in London?

>   RICK BLAINE
> When you get there, ask me!

>   CAPTAIN RENAULT
> Hmm! Diplomatist!

>   MAJOR STRASSER
> How about New York?

>   RICK BLAINE
> Well, there are certain sections of New York, Major
> that I wouldn't advise you to try to invade.

Too often, writers use dialogue to make points and to tell us things about the character and about that character's situation without the pithy wit of the *Casablanca* screenplay. When that happens, the film is dragged down to a talky mess that can alienate and bore the audience. And that's particularly true when the dialogue is in-your-face and obvious.

Some of that is seen in the films of Oliver Stone. Remember the "preachy" scene in *JFK* when Kevin Costner addresses the jury, or the preachy scene in *Wall Street* where we are once again hit over the head with the idea that Gordon Gekko is ruthless and mean? Those long-winded speeches from *JFK* and *Wall Street* could have been cut way down and the films they dragged down still would have made an impression. What I call the "Greed is good" speech (I quoted part of it in the "BAD" chapter) was delivered as an actual real-time speech to a group of shareholders and went on, nonstop, for two single-spaced dialogue-format pages! I can assure you that many a movie watcher's mind wandered during that bloated bit of writing!

We'll take a closer look at long speeches in an exercise at the end of this chapter.

Preachy dialogue was a particular characteristic of Hays Code-era movies. In older films, screenwriters were fond of having their characters make speeches—sometimes directly to the audience. That happened a lot (and still happens) in trial films but it also happened in films where the writers especially wanted to make a point. And in those cases preachy speeches were so obviously preachy that some of them were even delivered by priests.

Those old-time movies, besides making their point, were also long-winded because they were more dialogue-centered. And people liked high-falutin' sounding pretty talk. I admit that I myself am partial to rhetoric. I like nothing better than to hear someone say something that is inspiring and beautifully expressed, that tugs at the heart and thrills the intellect. I adore the chatty wit of screwball comedies and those old black and whites of the 30s and 40s where people talked with eloquence and near-English accents reminiscent of the British theater. But so what.

Today, those kinds of films drive audiences away. We've got short attention spans and even shorter hearing capacities. We're manic about pictures and we are too sophisticated and cynical for the altruistic art of yesteryear. Modern audiences probably wouldn't sit still for the politically talky lecturing of *Mr. Smith Goes to Washington* but they did sit still for *The Insider, The Contender,* and lots of other talky but political films because they were written in suspense–adventure styles.

Will audiences today sit still for movies with themes as obvious as American patriotism the way they did during World War II? They might if those movies are written in styles that sublimate message to action, intrigue and character. Perhaps "message" movies about the current state of world politics can only entertain audiences, if, as *Three Kings* did, they are made palatable by humor and bravado.

Right after the United States declared war on terrorism, September 11, 2001, the American government began meeting with screenwriters to pick their brains for "pre-emptive intelligence" for possible terrorist scenarios. Many of us were offended by this notion and some of us found it downright funny. Wasn't the American military creative enough to make up scenarios of its own? It all seems a little creepy to me now even though it seems relatively harmless.

Not so harmless, though, is the fact that the U.S. government has also met with filmmakers to talk to them about making films that would support American military action. That kind of involvement by government seems unethical in that it seeks to make art into propaganda and we've already talked about how odious watching propaganda can be. Of course, most audiences feel that religious propaganda is more problematic than the propaganda produced for its own citizens by a country at war.

Personally, I have no problem with patriotic films. I rather like them and certainly champion activities that make Americans confident and proud of their country. But, whereas patriotic films are important morale boosters in troubled and war-torn times, these films should be made by film makers who want to make them as an artistic statement and not because they are "encouraged" to do so by their government.

History has shown us the extreme to which government-sponsored filmmaking can go. Look at the movies made by the Third Reich's Ministry of Propaganda in Nazi Germany. To this day Leni Riefenstahl's *The Triumph of the Will*, a 1934 documentary about the 6th Nuremberg Party Congress, and her 1938 film of the XI Olympic Games in Berlin featuring Jesse Owens, are considered propaganda and not the art and brilliant example of innovative filmmaking Riefenstahl always maintained they were. Reifenstahl and other filmmakers in that era claimed that Hitler strong-armed them into making particular kinds of film and, today, those filmmakers are judged harshly.

I am certainly *not* equating U.S. government "encouragement" of patriotic films with Hitler's Ministry of Propaganda. But I do believe that "encouraging" patriotism can be just as unethical, as for example, "encouraging" networks to place obvious anti-drug messages in popular prime time television shows . When, in January 2000, Salon.com exposed the fact that the government gave television networks money for programs that contained anti-drug themes, creative people and others who found that practice unethical were up in arms. Their point? It's just as dangerous to dictate content as it is to limit it. As the famous American photographer Ansel Adams said, "No man has the right to dictate what other men should perceive, create or produce, but all should be encouraged to reveal themselves, their perceptions and emotions, and to build confidence in the creative spirit."[69]

That means if a screenwriter feels patriotic, that screenwriter will find a way to express that patriotism through characters and story, unencumbered by government interference. The danger in making films "encouraged" by the government is always that they might be-

---

[69]www.Samwed.com

come propaganda or considered to be propaganda and that's a label no artist wants to have applied to his or her work.

Simply by writing as an American and living in America, American screenwriters and filmmakers create films that reflect their particularly "American" point of view. That Americanism is evidenced in a style that is recognized around the world and expressed unconsciously by American screenwriters. American screenwriters who have been strongly influenced by European films, for example, may write films with a European look or feel but inevitably these films will still have an underlying American sensibility. For example, films by Woody Allen (strongly influenced by Italian cinema) are talky and introspective (like European films) but at the same time concern themselves with the neurosis, conflict, and pressure of typical New York (American) life.

American films do "preach" to audiences (particularly in foreign countries) almost in spite of themselves, about the American way of life. And that's evidenced by the ways in which unsophisticated foreign audiences sometimes see American films as documentaries about what it's like to live in America.

When I was working in the Arctic, the Inuit believed that American streets were fraught with the kind of violence depicted in crime shows—so much so that they dreaded setting foot in an American city and believed that every citizen was armed and often involved in shootouts! And when I taught master classes in Portugal, many of the students there believed that the violent American films they saw were religiously reflective of the American way of life. Of course, not all American subliminal film "propaganda" is negative. Because of American films, lots of foreigners (my Portuguese students included) also see America as an affluent, free, and fun place where anyone would love to live and where most everything is perfect. Certainly our romantic comedies tend to paint that kind of picture.

A screenwriter who doesn't want to "preach" needs to be careful about the environment of films as well as the dialogue in them. That means that certain "statements" about certain kinds of environments need to be carefully considered before they are created in a script. This brings us back to the issue of stereotyping. If a screenwriter stereotypes an environment (like the housing projects or the suburbs, for example) then that screenwriter is "preaching."

In Ang Lee's *Ice Storm* (1997, written by James Schamus), the suburbs were painted as a hotbed of immorality and seething dysfunction. This could be considered a stereotype and, in fact, the film was "preachy" because it made that statement. But the film's focus on character and the skill with which the movie was made softened its preachiness and made it successful. Characters didn't go on and on about the banality and boredom of suburban lives. The audience was simply allowed to see those things and react to them. There were no prolonged angst-filled speeches that made the film into a sermon about suburbia.

It bears repeating that all screenwriters who've studied their craft know that dialogue supports picture, picture doesn't support dialogue. What characters say must be brief and to the point. Dialogue can still be eloquent and literate. It just doesn't have to be long. Consider some of the good lines we remember: "Go ahead, make my day" (Harry Callahan, written by Joseph Stinson, *Sudden Impact*, 1983) and that great speech in *Dirty Harry* (1971, written by Harry Julian Fink and Dean Reisner):

> HARRY CALLAHAN
> I know what you're thinking. Did he fire six shots
> or only five? Well, to tell you the truth, in all this
> excitement, I've kinda lost track myself. But being
> as this is a .44 Magnum, the most powerful handgun
> in the world, and would blow your head clean off,
> you've got to ask yourself one question: Do I feel lucky?
> Well, do ya punk?

Personally, I thought the line about the .44 Magnum sounded forced and could have been left out. But even so, we get the character of Harry, and the point of the scene.

And remember the wonderful closing exchange between Ilsa and Rick on that airport runway in *Casablanca*? It "explained" to the audience and to Ilsa why Rick was staying behind and yet packed a powerful emotional wallop that did the screenwriters proud.

> RICK BLAINE
> Last night we said a great many things.
> You said I was to do the thinking for
> both of us. Well, I've done a lot of it since
> then, and it all adds up to one thing: you're
> getting on that plane with Victor where you
> belong.

> ILSA LUND
> But, Richard, no, I ... I ...

RICK BLAINE

Now, you've got to listen to me! You have
any idea what you'd have to look forward to
if you stayed here? Nine chances out of ten
we'd both wind up in a concentration camp.
Isn't that true, Louie?

CAPTAIN RENAULT

I'm afraid Major Strasser would insist.

ILSA LUND

You're saying this only to make me go.

RICK BLAINE

I'm saying it because it's true. Inside of us,
we both know you belong with Victor. You're
part of his work, the thing that keeps him going.
If that plane leaves the ground and you're not
with him, you'll regret it. Maybe not today. Maybe
not tomorrow, but soon and for the rest of your life.

ILSA LUND

But what about us?

RICK BLAINE

We'll always have Paris. We didn't have, we lost
it until you came to Casablanca. We got it back
last night.

ILSA LUND

When I said I would never leave you.

RICK BLAINE

And you never will. But I've got a job to do.
Where I'm going, you can't follow. What I've
got to do, you can't be any part of. Ilsa, I'm no
good at being noble, but it doesn't take much to
see that the problems of three little people don't

amount to a hill of beans in this crazy world.
Someday you'll understand that. Now, now ... Here's
looking at you, kid.

Lines like that can still move an audience to tears, even today. If you
can write dialogue like that you'll really be saying something!

Exercise

Here are two particularly preachy and overly long speeches from a
couple of old films. Your challenge is to cut these speeches down as
much as you can to still get the essence of the speech, keep what's
good in the writing, and stop short at preaching to the audience. Remember, this is only an exercise in getting the essence of dialogue
and reversing its preachiness. I'm not talking about the whole movie
here. Nor am I trying to "rewrite the masters." We're using these
speeches out of context as exercises.

Example #1

On the Waterfront (1954)
Written by: Bud Schulberg
Directed by: Elia Kazan

This film won eight Oscars, among them Best Screenplay and a WGA
screenwriter's award. It's about unions and corruption on the New York
City waterfront. The following speech is delivered by Father Barry (Karl
Malden), a priest who tries to stop the violence and racketeering by getting scruffy boxer Terry Malloy (played by Marlon Brando) to testify
against corrupt union bosses. Brando won an Oscar for his performance. Remember that great "I could have been a contender!" speech?
Kazan made the film after he named names in the McCarthy hearings.
Some say *On the Waterfront* was Kazan's apology for "ratting," but others from that period still haven't forgiven him. I'll run the speech without including descriptions of reactions to it that break it up on the page.
    (Father Barry) stands over the body of "Kayo" Nolan, which lies on
the pallet and has been covered by a tarpaulin.

FATHER BARRY

(Aroused)

I came down here to keep a promise. I
gave "Kayo" my word that if he stood up to
the mob I'd stand up with him all the way.
Now "Kayo" Nolan is dead. He was one of
those fellows who had the gift of getting up.
But this time they fixed him good—unless it
was an accident like Big Mac says.
Some people think the Crucifixion only took
place on Calvary. They better wise up. Taking
Joey Doyle's life to stop him from testifying is
a crucifixion—dropping a sling on "Kayo" Nolan
because he was ready to spill his guts tomorrow—that's
a crucifixion. Every time the mob puts the
the crusher on a good man—tries to stop him
from doing his duty as a citizen—it's a crucifixion.
And anybody who sits around and lets it happen
keeps silent about something he knows has
happened—shares the guilt of it just as much as
the Roman soldier who pierced the flesh of Our Lord
to see if He was dead. Boys this is my church. If
you don't think Christ is here on the waterfront,
you got another guess coming. And who do you
think He lines up with? Every morning when the
hiring boss blows his whistle, Jesus stands alongside
you in the shape-up. He sees why some of you
get passed over. He sees the family men worrying
about getting their rent and getting food in the
house for the wife and kids. He sees them selling
their souls to the mob for a day's pay. What does
Christ think of the easy money boys who do none
of the work and take all of the gravy? What does
He think of these fellows wearing hundred and fifty
dollar suits and diamond rings—on your union dues
and your kick-back money? How does He feel about

bloodsuckers picking up a longshoreman's work taband
grabbing twenty per cent interest at the end of
a week? How does He, who spoke up without fear
against every evil, feel about your silence?
You want to know what's wrong with our waterfront?
It's love of a lousy buck. It's making love of a buck
—the cushy job—more important than the love of a
man. It's forgetting that every fellow down here is
your brother in Christ. But remember, fellows, Christ
is always with you—Christ is in the shape-up, He's in
the hatch—He's in the union hall—He's kneeling here
beside Nolan and He's saying with all of you—if you
do it to the least of mine, you do it to me! What they
did to Joey, what they did to Nolan, they're doing to
you. And you. And you. And only you, with God's help
have the power to knock'em off for good!
(Turns to Nolan's corpse)
Okay, "Kayo"?
(Then looks up and says harshly)
Amen.

Example #2

Knock on Any Door (1949)
Written by: Daniel Taradash and John Monks, Jr.
Directed by: Nicholas Ray

This was the first film produced by Humphrey Bogart's own production company, Santana Pictures. In it Bogart plays Andrew Morton, a tough lawyer who takes on the defense of a street kid named
Nick Romano who's charged with the murder of a cop. It's a trial
film filled with flashbacks of the kid's hoodlum life. This speech is
from Morton's summation to the jury after Nick breaks down on the
stand and confesses to the murder Morton thought he didn't commit. It's almost a synopsis of everything that happens in the film.
Again, I'm not going to split the speech up by the business of court-
room reactions.

MORTON

Your honor, there's something I'd like to say
on behalf of the defendant. When I took this
case I believed Nick Romano was innocent.
I believed what he told me and I believed what
those men Butch and Sunshine told me. I believed
because I wanted to believe that all the filth, fury
and jumble of this boy's past has not produced a
killer but Nick Romano is guilty. He's guilty of many
things. He's guilty of knowing his father died in prison.
He's guilty of having been reared in poverty. He's
guilty of having lived in the slums. He's guilty of
having the wrong companions, the pickpockets
and hoodlums, panhandlers and prostitutes of
the worst district that ever disgraced a modern city.
He's guilty of the pool rooms and bars that were
open to him as a boy. He's guilty of the foul treatment
of a primitive reform school. Keep the boy and
civilize the boyhood. This boy could have been
exalted instead of degraded. Student instead of
savage. But come with Nick on his own way to that
reform school. See your best friend die of subhuman
punishment. Come with Nick along skid row where the
fences buy anything and with no questions asked.
Come with Nick into the alleys and on the streets
 past the drunks and panhandlers and prostitutes
into the pool rooms and the bars. Listen. Listen to
the jackrollers and the thieves, absorb their poisonous
philosophy of life. Come with Nick to the penitentiary.
Be numbered and counted and hated and leave there
determined to be worthy of that hatred. Yes Nick Romano
is guilty but so are we and so is that precious thing
called society. Society is you and you and you and all
of us. We, society are hard and selfish and stupid. We are
scandalized by environment and call it crime. We de-
nounce Crime and yet we disclaim any responsibility for

it and we lack the will to do anything about it. Until we
do away with the type of neighborhood that produced
this boy, ten will spring up to take his place—a hundred
thousand until we wipe out the slums and rebuild them.
Knock on any door and you may find Nick Romano.
The newspapers have been clamoring for pictures
and stories about this trial. But why don't you print
this? They and you and I, we the good people,we the solid
citizens of this community, we photo-
graphed and labeled this boy years ago. We made
this rendezvous with him years ago. We brutalized
and ordered him here years ago. If he dies in the
electric chair we killed him. Print that.
(To the Judge)
Where do we take him now? Do we kill him?
The current coursing through the blood, the nerves,
the heart, the brain. I ask mercy from this court. I
ask this so that for us who walk free, for all of us,
there may be some mercy.

In spite of Morton's speech to the jury, Nick still fries.

It's not hard to tell from these speeches what messages the writers
were trying to get across. Fortunately, both films were good enough to
be successful in spite of their belligerent preaching. That's because
the acting in both films was great and because both films were written by screenwriters who had good ideas and weren't afraid to use
them. Sometimes, besides talent, integrity, and good sense, you also
need good ideas.

PART VI

HAVING WRITTEN
AND WRITING MORE

WHAT'S THE IDEA?

'm always amazed when people say they want to be writers or that they are writers, and then tell me they can't think of things to write about. It seems to me that part of writing is having so many ideas about life, about people, places, and events, that there could never ever possibly be a problem thinking of stories to tell. But, of course, I realize, too, that having too many ideas can cause some confusion. Sometimes, writers can be overwhelmed (or underwhelmed) by all the things there are to say.

That's when personal value systems play an important role. Strong values give writers strong points of view and strong points of view make for strong approaches to subjects. Now that you've discovered your personal value system, point of view, and ethical framework, you're ready to test these things out on your own ideas. Where do you find your own ideas? All around you. The best ideas come from your own real life and, as we've all heard ad nauseam, real life is often stranger than any fiction. To learn how to take ideas from real life and make them your own, do the following exercise.

Exercise

1. Read at least one newspaper each morning.
Choose newspapers rather than radio or television because most newspapers deal with the events of the day in more depth. Of course, the better the newspaper, the deeper the story. I like to read at least two newspapers a day—my local paper and *The New York Times*. If I had more time, I'd probably read *The Wall Street Journal*, and *The Christian Science Monitor*, too.

When possible also read the following: one scientific magazine or journal, one magazine devoted to health issues, a travel magazine, a political magazine, a magazine about people (*Biography* or

People, *US*, etc.), and finally, a magazine pertaining to your religion and/or personal belief system.

2. As you're reading, notice the kinds of things that affect you—the stories that bring out an emotional response. For example, in one morning's paper I was stirred by the following stories:

> The funeral of a couple and four of their eight children killed when a terrorist bomb went off in a crowded Israeli pizzeria. One of their surviving daughters, seriously injured, attended the funeral on a stretcher.

> The rescue of a mentally disabled woman from a small Eastern European village where she was kept locked up in a cowshed by her parents for over 20 years. Everyone in the village knew that the girl was there but said nothing.

Each of these stories brought up certain value issues for me. In the newspaper you read this morning, make a list of the stories that struck a chord in you and make notes on them.

3. As you review each of your story notes, write down the value issues that surface for you. In my case these were:

    a. Is terrorism a political/military act when it harms innocent civilians?

       • How does someone emotionally survive the death of most of his or her family?

       • Is revenge ever justified?

    b. If an entire village knew that the mentally disabled woman was being kept prisoner, why didn't someone speak up to free her?

       • Did her situation cause friction between some of the villagers?

       • How did her parents come to the conclusion to imprison her and how did they live day-to-day with this decision? Did she have brothers and sisters who did?

       • How was she finally rescued?

4. Now, go deeper into each story and imagine one person (a character) who is affected by the events of that story. The person doesn't have to be a main participant in the event. He or she can be an observer or someone who seems inconsequential.

5. Write down how the events in the newspaper story might impact the life of the character you have imagined.

6. Create a biography for that character and a value system. (You can use the list of values we came up with in an earlier chapter and/or others you have already determined for yourself.)

7. Insert the character you created into the events of the news story. At each crucial story point, write down what decision the character will make based on that character's value system.

8. You can, if you like, use the event as a jumping-off point that takes your character in a whole new direction and has that character come to a new realization about life and events, or, you can simply insert your character in the events as they unfold. Choose the approach that is most exciting to you.

For example, in my terrorism story:

My character is the teenage survivor. Her values are: a devout belief in God (she is a practicing Jew), a reverence for life and a belief that any killing is terribly wrong, that conscience must guide everyone in life, intense Israeli patriotism, a deep love for and a profound loyalty to her family, a desire to be free from incapacitating grief, the belief that terrorism is an illegal, unethical, and immoral act, that honesty is profoundly important, that success comes from the ability to live out her value system.

My story is about how she comes to terms with the death of her family by making a decision to avenge them and by compromising some of her values to do that. Here's a quick thumbnail story outline broken into acts.

Act One: Intro to the environment of the country and its political situation. Intro to the girl, her family, and her values. The incident occurs. She is inconsolable and then through a process of discussion, research, and information gathering and with the "help" of "experts" (Bok Model), she makes the decision to avenge her family.

Act Two: She joins a militant Israeli anti-Palestinian group who asks her to infiltrate the family of a suspected terrorist group. In doing so, the girl must compromise more of her values (she has to stop practicing her Jewish rituals, she has to lie and deceive).

In her undercover work, she learns more about the charged political situation in her country, experiences firsthand the complicated emotions on both sides, and then must make a decision to kill the people she has grown to understand, or to seek another way of coming to terms with the loss of her family.

Of course, it's possible to throw in a little Romeo and Juliet love affair here but I wouldn't do that because it's too clichéd. Rather I think I'd have her falling for the head of the anti-Palestinian force who gets her to infiltrate and whom she must risk displeasing if she holds fast to any of her earlier values.

Act Three: The resolution of actions she takes in Act Two. I'll come up with that later.

You can see how I've been able to build in some complicated psychological issues here and how I've set the story up for ethical decision making that would create suspense and inner conflict. And I've just come up with this story. I didn't know it before I read the paper and I was able to flush out this general idea in about 20 minutes.

9. Now you try it. Use a story you've found in this morning's newspaper to come up with your own story and a thumbnail act outline.

10. Tomorrow, and every day for the next week, read the paper and then make yourself come up with five story ideas based on things you've read. You might even surprise yourself and come up with many more than five. And remember, because you are just using the event that happened and fictionalizing the characters, you also get to fictionalize parts of the event if you want to. That way you don't need the rights to the story. You are only using the "real story" to fire up and fuel your imagination so it can take off and create full-blown scenarios that might have no similarities to the original newspaper story.

In my terrorism story, for example, I can change the location of the bombing (a car dealership instead of a pizzeria), the number of siblings killed, even the ages of the survivor and her parents. I can, if I like, even change the locale from Israel to Ireland, and instead of the Israeli–Palestinian situation, I can do a Protestant–Catholic, Irish situation, or any other racially or ethically charged world location. Unfortunately, there's no shortage of those.

Earlier, we talked about writing as a process of self-discovery. In fabricating stories, if we relax and let our intuition take over, we may find the story we create unfolds so naturally that we don't realize until later that it's bringing up hot button issues that have direct relevance to our own lives. It works almost like magic. And often, it's a better idea to let ourselves naturally be drawn to subjects by feeling than it is to look for subjects specifically because of issues we're interested in addressing.

We can, though, work from personal life issues if we want to. Here's how.

11. Write a character biography of yourself. Make it as objective as possible. It might help you to see if there's a pattern you've been developing. It will reveal to you your interests and proclivities.

- Create a Choice Chart of your own life. Write down (in order of importance to you) the decisions you made that caused your life to go in the direction it has.
- Write down some of the things that might have happened to you if you'd made different choices.

This personal technique should provide a great amount of material for you. Remember, when you use your own life, you might also involve the lives of those close to you. It is a good idea when you do that to fictionalize those people as much as possible. You can even fictionalize yourself. Changing those you know will afford you new opportunities to experiment with true facts and turn them into fiction. It will also keep those you love from feeling exploited. Remember that we're not making documentaries here. We're writing movies that spring from imagination although they may have a toehold in fact.

Other People's Lives: Free Ideas?

Whereas it's okay to use news and real events to cull ideas for fictional stories, it's unethical to use other people's life stories and situations *exactly as they happened* unless you ask their permission and option those stories. (We'll talk about how to get the rights to someone's story later.)

You can use historical stories that are in the public domain (happening at least 75 years ago) exactly as they happened if they were reported, but you can't use a person's life if that person is living. If that person is dead, you might also need permission of heirs. Better to keep to public records.

As for your personal family and friends, you may think that your wacky sister or your bad-luck best friend would love to have a movie written about them, but as enthusiastic as they may seem during the writing, things can turn ugly once money comes into the picture or a screenplay is produced. If you're writing a movie about your own life, you have the rights to do that as long as you make clear that it's from your point of view. We all, fortunately, have the rights to our own life stories. And you can always use one true event from your own life and fictionalize everything around it. You need to be bold and improvise.

Here's a little Event Exercise that will give you even more ideas.

Event Exercise

Make a list of particularly memorable or interesting events in your own life. Wrap stories around each event or use each event to kick off stories not related at all to your own life.

Here's a personal example. My first movie was about something that actually happened to me and then I expanded that into a whole movie.

When I was about 8 years old, my mother sent me to the Chinese laundry around the corner to pick up shirts. I'd never been there alone and usually waited outside while my mother went in because I didn't like the smell of steam and starch that came from the open door. That day, as I entered, the smell was stronger than usual and was accompanied by the sound of a tinkling glass wind chime. The tiny shop was empty.

I was about to call out when suddenly, from behind a slit in a gold brocade curtain behind the counter, a wizened yellow claw-like hand with overly long vermilion fingernails poked toward me. I stepped back. The claw was followed by an arm draped in brown linen. I moved even farther back. And then I realized that the arm was attached to a little old Chinese lady—wrinkled as a prune and so decrepit she seemed to lurch when she moved. I gasped. She was the first old Asian person I had ever seen. Terrified, I ran out of the shop and all the way back home.

Years later, I wrote *Home Free*, a kid's movie that began with the Chinese laundry incident and then went on to explore racial prejudice that stemmed from fear of unusual people. In my movie, a 9-year-old Caucasian girl moves into a predominately Asian neighborhood, visits a Chinese laundry, and has my "real event" experience. But then, I move into fiction as my main character starts school and is invited to a birthday party where she's the only White person. Even worse, the grandmother of the birthday girl is the terrifying apparition at the Chinese laundry.

As the party progresses, the Caucasian girl feels more and more out of place. Her differences from the Asian group create awkwardness and self-conscious mistrust—until the grandmother helps her out during a game of hide and seek. By this interaction and simple act of acceptance by the once-feared grandmother, the Caucasian girl relaxes, and opens herself up to having fun with the other kids. She leaves the party understanding how it feels to be an outsider and realizing that her fear of another culture came from her reluctance to engage individuals from that culture and to have meaningful interaction with them.

In real life, I never did see the old woman again and I stayed away from the Chinese laundry, but I did become fascinated with Asian culture and was able to work out the effect the laundry incident had on me in a way that eventually led to a screenplay with a message.

Home Free won funding, distribution, and ultimately, national and international awards because it achieved its goal of promoting multi-culturalism. And it all started with something that really happened to me when I was a kid. Sometimes, we may think that the mundane events of our own lives are insignificant, but if we look at them with imagination and energy, we might find that they can be jumping off points for wonderful stories that can enrich us.

It's helpful, for that reason, to keep a journal or diary in which you record not only things that happen but also the emotions you experience when they do. Often, simple things can take on great meaning by the energy that we invest in them, and one of the strongest forms of energy is emotion. I'm not talking about hypersensitivity, sucky thoughts, or cloying ways of being. I don't think that your entire life has to be one big empathy card. What I'm talking about is really being present in your own life so that you always notice what is happening to you and around you.

This is not a new concept. Mystics talk about this way of being as "mindfulness" and being "in the present." What that means is that you give full attention to things that are happening around you as they happen without being absent-minded. That kind of attention is invaluable to a writer. It doesn't mean reading anything extra into every event but it does mean close observation of every event as it happens.

This is a skill newspaper reporters use to do their jobs well. Unlike television reporters, newspaper reporters don't have cameras along with them to record the action. Instead, they have to train themselves to really take in the events they're reporting. I learned this the hard way when I was a "cub" on *The Toronto Star*.

When I first started working for the paper as an intern, I was a freshman in college. I had big reporter dreams. I wanted to be like the girl reporters in *Front Page* and all those other 30s and 40s movies I saw at the Adelphi, so I wore hats (can you imagine an 18-year-old in hats!) and high-heeled shoes even at the rewrite desk.

My first assignment was to cover a high school production of *Macbeth* put on to raise funds for the Red Cross. I spent 18 hours writing three witty paragraphs (I still remember my lead: "Shakespeare bled for the Red Cross yesterday"!) and was overjoyed when my story made the front of the metro section (in a box). As I sat on the subway on my way to work the morning the story came out, I watched people reading the paper and glowed with delight.

I was "rewarded" for that success by the entertainment editor who sent me out on what might be a real "color piece." I was to cover the arrival at a large outdoor theater (at that time the largest stage in the world) by Stanley Holloway, an old vaudevillian and British actor who was coming to our city to do a show.

As a lowly cub, I wasn't allowed to interview Holloway. I was only supposed to follow him around and describe his reactions to the enormous theater. I spent about 20 minutes with him as he toured the place and then I went back to the paper and wrote what I thought was a sensitive and engaging piece about an old actor being overwhelmed by a big stage. The story (four paragraphs) took me all night.

The next day, right on deadline, I proudly brought my copy to the entertainment editor. He was a gruff former crime reporter with a yen for showbiz who thought women didn't belong in city rooms and that college students were lazy, shiftless, rich brats. He took one look at my story and threw it back in my face. Then he started yelling. "Were you even there?" he thundered. Apparently it wasn't clear from what I had written.

And then for the next 5 minutes (it seemed like an hour) he ruthlessly grilled me with questions about minute details I had missed in my eagerness to wax lyrical. "What color shoes was Mr. Holloway wearing? What kind of socks? What kind of tie? What kind of jacket? Did he smoke a cigarette? Did he wear a watch? What kind of shirt did he have on? How was his hair combed?" On and on until I was in tears .

When his tirade was over, I retreated to the nearest bathroom. As I sobbed in a cubicle, cursing that editor and damning him to an obscure job on the rewrite desk of the Fleasgroin Gazette, I grudgingly admitted to myself that even though he was a jerk, he was also a little bit right. I had missed the details. I hadn't really been present enough to notice the little things that make for subtext in a story. I promised myself that I'd never go out on a story again without seeing everything and being 100% there.

This resolution has helped me many times in professional and in private life. I found as a reporter I was able to remember conversations word for word (important for quotes) and to relive the events so clearly in my own mind that I could relate them colorfully and compellingly later on. I also noticed that developing a strong power of observation helped me immensely in my screenwriting. I am able to describe things succinctly and efficiently and yet evocatively as a result of my experience with present-mindedness. I also find that observing carefully does in fact make you more involved in the scene rather than less involved. It improves your listening skills (a big help in writing dialogue) and makes you more sensitive to others. People are always saying that observers are indifferent or casual. Not at all. If you profoundly observe, you can profoundly experience. The key is intensity. Taking that approach then, your life will appear richer, fuller, and more compelling, and the events that happen to you will make for much better screenplay material.

Observation Exercise

Think of what you were doing last night. Write down what you can remember about what happened in as much detail as possible. Try to remember the sounds, sights, smells, colors—everything. This could work even if you were at home in your chair watching TV.

Tonight, plan to spend the evening carefully observing as you are going about an activity. Do not make notes.

- Tomorrow morning, write down what you can remember about what happened in as much detail as possible.
- Compare the two records (one before conscious observation and one after). Also compare how well you remembered the second event as compared to the first.

At first, conscious observation may seem tiring but eventually, if you make it a habit, it will become second nature to you. This is a new way of thinking and being and it might take time to incorporate into your personality, particularly if you are used to going around "spaced out" and oblivious, self-absorbed, or lost in thought.

Observation, incidentally, will also help you with your concentration during the writing process. You'll be more aware of what you are writing and how you are writing it and that might save you from having to labor over every word and reworking things compulsively.

You might also notice that it will help you "read" people. If you are a careful observer, you might also become a better judge of people because you'll notice their subtle gestures, movements, and facial expressions indicating attitudes, moods, and inclinations. This will help you to create compelling characters and stories that connect.

ALL'S FAIR IN LOVE, WAR, AND SHOWBIZ?

an you blame anyone for wondering? Everyone in showbiz is so desperate and determined to make it that the temptations are there to die or kill trying. Screenwriters are always talking about "ethics" but in a roundabout way. The conversation usually goes something like this: "I was screwed on that deal" or "He really screwed me over" or "He got screwed on that one" or "I screwed him before he could screw me" or "He screwed me before he screwed himself" or "He's screwing himself and doesn't know it" or "He's screwing me and every-one knows it." The unofficial anthem of showbiz might be Bob Dylan singing "Everybody must get screwed."

I know it sounds a little like I'm channeling George Carlin, but with all this screwing going on—screwees and screwers, the nearly screwed and the newly screwed—nobody has time to actually write. So if you want to write, keep it simple. Remember, screwing gets done in the biz, not in the writing. Oh, you can have people "screwing" and you can even have your characters screwing and getting screwed, but whether or not you will be screwed as a writer shouldn't make one bit of difference to you.

What others can or cannot do to you, and what they do and have done to you, doesn't matter in the least. Even if you get ripped off so many times you have to change your middle name to Velcro, take it on the chin and move on. Naturally, because screenwriters have soft hearts and long memories, you will feel the pain, but if you want to work in the industry you can't let that pain get to you and cripple you.

When people say that you've got to be persistent, this is what they mean. We can all take a tip from actors. They get rejected over and over and over again and still they dress up and go to auditions. And like them, you've got to keep going until you decide that it's not worth it any more and you just can't take any more pain. That point of knowing when to quit is different for everyone.

Super-writer David Mamet, who's quit being a screenwriter-for-hire, was asked if there was an experience that made him say, "Enough."

214

"Oh, sure, every experience," he said. "On Hannibal, they ended up not using my script at all. I thought it was a pretty good script. That was a little disturbing. But it's like you're an architect and you build some rich people a house. Then they invite you over for a cocktail party, and you say, "Those pigs, they moved the sofa. How dare they?" But that's what they paid you for, whore that you are. They paid you to nod and say, "OK, you've bought my right to complain.""[70]

Even David Mamet puts up with pain. In spite of it, he's still writing screenplays, albeit in a different way and with an eye to direct them. Clearly, he's not yet hit his ultimate pain threshold that would make him give up writing movies all together. Some people never do. But believe me, you'll know when you do. One morning you'll wake up and you'll just know. But until that time, you've got to maintain a positive attitude, develop a thick skin, and know that you are bound to get screwed sometime and there is very little you can do about it.

I say these things not to discourage you or frighten you. If you truly "gotta" write (and early on in this book you were given the opportunity to think that through), then nothing anyone says about pain and show business horror will stop you. If you "gotta" write movies, then you're going to do it no matter what. But if all this talk of screwing and getting screwed has you worried, then you might want to go back and reassess your motives for writing. It's not too late to do that! In fact, it's never ever too late to do that.

It might make you feel better to know that the Writer's Guild does have a bunch of regulations that try to prevent writers from getting screwed. Unfortunately, cunning and wily screwers will know ways to get around these little protections. The reality is that the more savvy the screwer, the better he (or she) screws. So the only way you can really protect yourself is by behaving impeccably, resist screwing anyone, and seeing how far integrity and ethical behavior can get you.

Because people are known by their reputations, it will soon get around that you are to be trusted and a joy to work with. There are good people out there, even though they are hard to find and perhaps, if you are vigilant, like will attract like. At least that's what I hope even though I can't promise it.

When young writers fear Hollywood, they do so because they believe that it is an unethical and dangerous place that might ultimately destroy them. Not true. Although it is true that screenwriting is a very difficult and treacherous thing to do if you want to earn your living at it, you've got to know that most legitimate people in the industry are

---

[70]Gordon, Devin. *Mamet's On A Classic Caper,* Newsweek Magazine, November 19, 2001, p. 69.

ethical and do not rip people off or try to steal from them. Legitimate companies want good projects and they aren't willing to steal or be shady in order to get them. That's because nobody wants a cloud hanging over his project—especially when lots of money and careers are at stake.

Reputable agents, production companies, and studios will get you to sign Submission Release Forms protecting them from possible lawsuits just to avoid problems. These release forms are so scary they are sure to frighten even the most seasoned writers. They read like licenses to rip off. Some sample stipulations: If anyone at the company comes up with an idea that's identical to the submission, it's a coincidence, or if anyone who used to work for that particular company comes up with a script identical to the submission in every way, it's a coincidence. Pretty deadly. It's not a good idea to sign any of these release forms unless you are willing to walk away from your screenplays.

If you submit your screenplays through a reputable agent, you won't have to. Unfortunately, agents make you sign release forms, too. You've got to trust that reputable agents have better things to do than to rip off good script ideas from people they could make money representing.

Young writers who complain they've been ripped off have been dealing with fly-by-night or untested producers who have no track record and promise big contacts and payoffs for free options that might tie up scripts sometimes for years. I do recognize that producers do have to start somewhere and sometimes it's okay to go with fledglings who can grow with their writers, but if you do decide to go with one of these fledgling producers, make sure that you don't sign an exclusive contract or that you tie your scripts up for more than 6 months at a time.

I know of one new and wonderful writer who made the mistake of tying up a viable project he should be using for a writing sample for a year and a half. The producer who has it refuses to let it go even though he's making no headway, can't get any meetings, and is too busy working his day job to hustle his movie-making dream.

Before you tie up your projects, make sure that you are given some assurance of meetings and contacts and that you will be invited to pitches. You've got to get something in return for trusting those who are new at the game. And before you do trust them, spend some time with them. Find out if they really are people you want to be associated with. Find out their motivations and priorities. Find out what drives them. Don't make bargains with devils and later bemoan your fate!

One sure thing, however, is that if you yourself are ethical and behave with integrity, you will be able to sleep nights and, in general, have peace of mind, and we've already established how important that is. So going in, there are perhaps some things you should know about ethical business behavior that will stand you in good stead and will make you look professional. That way at least, when you get screwed, you can sue.

AGREEMENTS, CONTRACTS, AND PITCHES

People are fond of saying that deals are about "relationships" and people hire their friends. Don't believe it. Sadly, show business strains friendships like sumos strain spandex. In business, when lots of money and careers are at stake, people can't afford to carry dead weights or make allowances for airheads even when those weights and heads belong to people they say they love. The best way to protect yourself and make sure that everything is ethical and clean is to put your business relationship in writing. Do nothing based on a handshake, a kiss, a wink, or even a blood oath. Get it in writing and get it evaluated by someone who knows entertainment law. Forget your cousin Blima who's a shark with escrow savvy. Entertainment agreements and contracts are so complicated and specialized they can really only be understood by people used to working with them. So hire a professional entertainment attorney or beg a favor from an entertainment lawyer who believes in you and is willing to invest in your career by agreeing to take 5% to 10% of what you get if and when you sell the project.

Even if you've decided to partner up with someone you trust implicitly (your best friend, your lover, the guy who changes your oil) make sure that you have a Writer's Collaboration Agreement. If you call the Writers Guild of America West in Los Angeles, someone in the Contract Department will fax a copy over to you. Fine print will tell you that the provisions in that contract "are not mandatory, and may be modified for the specific needs of the Parties, subject to minimum requirements of the Writers' Guild Basic Agreement."

Get legal advice and amend that agreement any way you want to before you sign it. The Collaboration Agreement stipulates ownership of the work, registration of that work with the Writers Guild, completion date, additional functions performed in regard to the work, and agent representation. Sign it and then get it notarized. You can never be too careful. If everything is spelled out beforehand, it makes the relationship much clearer and easier to work in. And your relationship will also be clear to anyone who wants to split you up. You can always point to the agreement and get yourself off many a hook. That way, you don't have to wait until the last minute to solve problems that could become serious deal breakers in a happy crunch.

That takes care of the legal problems. Now what about the actual ethical dilemmas that come up within the writing partnership? These are much harder to solve, so you might want to work them out before you start. For example, how much work should each person do? And what steps will be taken if suddenly, in the middle of the project, one partner finds that he or she is doing all the work?

Be very careful to talk out all the possibilities. Even though it might be unpleasant to imagine hideous biting and scratching scenarios that might be part of your partnership future, it's essential to do that to avoid broken hearts. You can minimize the unpleasantness by using your creativity skills to make up daunting problems. And if you do this, you might even get some great screenplay ideas out of your interaction.

Along with the interpersonal scenarios, also determine who will get credit for the story idea and for writing the screen play, who will be present at pitches, who will do the talking at meetings, who will make the phone calls and do the photocopying. All of this may seem trivial, but believe me, in the heat of writing and especially in the frost of selling, it's important to make sure there is no resentment building up because someone thinks he or she is giving all the guts and getting none of the glory. Look on the Collaboration Agreement as a pre-nup and things will go more smoothly for you.

For writers who trust too much, here's a personal Horror Vault example of what happened to me one time when I didn't have a Collaboration Agreement. About 12 years ago, I was close to a graduate student I'll call Lafcadio. After he graduated, we kept in touch for a while and then lost contact. About 7 or 8 years later, he called out of the blue. He told me he had some success writing television shows and I was thrilled for him. We chatted a while and then he asked me if, as a favor to him, I'd partner up with him to write for a television show to which he had access.

The show was shot in Canada and required that some writers be Canadian. Lafcadio reasoned that because I was a Canadian expat, he'd have a better chance of getting an assignment if I teamed up with him. He said that we had a small pitch window and we needed to come up with something in a few days. Reluctantly, I agreed to be "used" in this way because it sounded like a quick and relatively painless process and I believed we might have a good chance of selling something.

I told Lafcadio that I'd enter into this partnership *only* if it were a serious work relationship: Both of us would have to be involved in the creation, the writing, and a face-to-face pitch of the show. He agreed, and because I liked and trusted him and time was short, we went to work without anything in writing.

The next day, we had a big brunch and came up with a story line. Then, we agreed that each of us would write a treatment of our idea that night, fax our treatments to each other, and hammer out a combo before the end of the week when we were scheduled to pitch to the exec in charge. I rushed home after our meeting and wrote my treatment. Late that night I called Lafcadio and told him I was ready to fax. He yawned. He wasn't ready. He had been watching TV with a friend and would have his treatment by the next morning. I went to bed.

The next morning there was no treatment. But that night there was an e-mail telling me that I had done far too much work, more than was necessary, and he wasn't sure he liked the idea after all. He'd write something that night and get back to me. I waited. No fax. I called. No fax. I called again and left messages. Days went by. I e-mailed.

On the day of our appointed pitch, I got an e-mail telling me he'd been able to get a 2-week pitch postponement and would work on the story and get it to me in a few days. I waited. Still no fax. I phoned. I e-mailed. Still no fax. Finally 2 weeks went by. I assumed we'd lost the pitch window. I phoned. I e-mailed. Still nothing.

Finally, 5 weeks later, he called to say hello. Hello? What about the project? "Didn't I tell you," he said. " I pitched it on the phone and the exec passed?" I went nuts. What did he pitch? My version? His no-show version? Some mysterious hybrid? And why wasn't I brought in on a face-to-face pitch (one of my conditions for working with him)? I was steaming mad. Lafcadio couldn't understand what all the fuss was about. He thought I was taking the whole thing too seriously. I tried to explain to him how unethically he behaved but it fell on deaf ears. Even to this day he maintains that I overreacted.

I soothed myself by being happy that I only lost 2 days of work instead of wasting a month on the project. But I'm still the loser in this. Because I don't have anything in writing, I don't feel comfortable running with the idea on my own or even turning it into a feature, because the nub of it was concocted during our big-brunch discussion. Even though ideas and concepts, if not written down, are not protected by copyright, I feel it would be unethical to claim the sole authorship of that one. The lesson in all of this? Before you do anything, even if you're pressed for time, and even if you trust and harbor great fondness for the little tyke you're working with, get a Collaboration Agreement.

Agreements will also protect you in projects using less ephemeral ideas. For example, if you want to write a script based on a book or someone's life story, make sure that you legally secure the rights.

With books, the process is pretty straightforward. Find a book you like, call the publisher, ask who owns the screen rights, and then write a letter asking that entity for an option to write and sell a screenplay based on the book. (An option is the exclusive right to a book for a period of 6 to 18 months, sometimes renewable.) If the book is a best seller or by a big name, expect to pay lots and lots of money for the option (even when it's against a steep final selling price that might be paid by a studio or network once they buy your script). Know that big books and books by popular writers are often purchased while they're still in galley form by producers and studios that have entire departments devoted to acquiring literary properties.

Your best bet would be to seek out those obscure little literary gems that are out of print or largely undiscovered. If you do that, here's what you need to do to be ethical. You need to call the publisher and ask who owns the screen rights just as you would with a big book. If you are lucky, the writer will own the rights. In those cases, you'll be referred to the writer's agent (or, if there is no agent, to the writer). Write the writer directly and ask for the rights.

When you do that, it's important to be personal and friendly. Don't come on like a big production company. Be honest and straightforward. Let the writer know how much you enjoyed the book and that you'd like a chance to discuss its cinematic possibilities. Also let the writer know that you too are a writer and not a big company and don't have much money.

It's all right to ask for a small-amount option if you make sure your final selling price is reasonable. Make sure that it's reasonable to you as well as to the writer of the book. Don't offer an eventual selling price that is so high it will be a deal breaker to any production company that may agree to make a movie out of the book. But make sure it's not so low that the writer of the book feels ripped off. Also, although some people pride themselves in being able to get free options, I feel it's always more respectful to offer the book writer a little something just as a show of good faith. Even $150 for 6 months will go a long way to building a relationship of trust between you and the book writer.

Make sure that you get an entertainment lawyer to draw up the agreement and notarize its signature. Once you have the rights to a book, then you may approach producers even if you don't have the screenplay adaptation ready. Sometimes all you need is the book, a brief synopsis, and a great pitch to get a development deal. But remember, *under no circumstances should you pretend that you have the rights to a book if you do not!* Some writers think that if they write a dynamite adaptation of a book, they'll be able to convince a studio or production company to buy it and then try to get the rights. That's a waste of time. No company would put out good money unless it was certain that it was working on a sure thing. And if it comes out that you lied, you're toast.

That's also true if you turn in a screenplay based on a real-life story to which you don't have the rights. If the story is an historical one or in the public domain, then you can do what you like with it. Stories that happened 75 years ago or to famous people and reported in the press, are fair game unless they libel people still living. Stories like that are usually a hard sell because they've already been thought of by lots of other people. The more desirable and valuable stories are those that are obscure, and happened to real people to whom an audience can relate.

In an earlier chapter, we touched on using life stories of the near, dear, and even distant. If you want to use someone's story, you've got to legalize your use by getting consent. Sometimes that consent doesn't come cheap. In life stories that are particularly prominent or notorious, you are going to have to pay lots of money. Just like best sellers, celebrity stories or stories about big shots are snapped up by big companies who troll for them by closely scrutinizing the press. If you want to make your mark, get the rights to a story by a person no one has ever heard of—a story with dramatic impact and universal appeal because of what it says about people and about life.

Once you meet someone whose story interests you, you need to draw up an agreement giving you the rights to that story. Just like with a book, you can option the story against a final selling price and get a lawyer to draw up the agreement. And just as in a book option, make certain that the person whose life story you are buying knows that the final selling price won't be humongous. People who think they have interesting life stories always think that they can make millions. The truth is they'll probably make only thousands, maybe $45,000, tops. Most people get $25,000 to $30,000 for their life stories. That's not a heck of a lot when you consider what they risk.

Here's the tricky part. If you want to be entirely ethical, you will have to let people who sell you their life stories know what they are in for. These people will have to realize that when they sell their stories, they also sell the rights to their name (if their real name is used), their likeness, and their privacy. Their stories can be changed in any way the moviemaker decides to change them. Usually, they have no say in the change. They have no input in casting or in anything else. People are always saying "I've got a great story and I can just see Kevin Costner playing me." Ha! They'll be lucky if they get Drew Carey playing them! They have to know that if they want to get their movie made, they're going to have to be willing to give up everything for it—and that sometimes even includes their self-image and self-respect. Surprisingly, more people are willing to do that than you may think. Take as evidence the madcaps that appear on Jerry Springer or Ricki Lake!

Of course, ethical screenwriters might have some problems with letting people they care about be trashed in the moviemaking process. Here's another example from the personal Horror Vault.

Many years ago, while hanging out at the gym, I met a woman (I'll call her Zelda) who was a champion professional body builder. Because I was into body building myself at the time, we got to talking and I found her life story fascinating. After a few months, I got to know her pretty well and she suggested to me that she always had wanted a movie made about her life.

The angle? A true story! A woman who is a fine specimen of good health, who wins body building prizes, and who goes around the country lecturing on nutrition and health foods is into drugs and bulimic! The woman overcomes these problems and wants to inspire other body-conscious women to conquer their demons and to take up body building in a healthy way.

The timing was tenuous. Bulimia wasn't a household word in those days. In fact, nobody was really talking about it, certainly not on television. This would give me the chance of educating the public about what even then was a craze among the body conscious and at the same time bring notoriety to a woman whom I thought deserved it. We drew up an agreement for the rights to her story. The agreement allowed me to make any changes necessary in the story to sell it.

I created a proposal and went in to pitch to a major production company. The pitch went well. The executive in charge of development was a woman who understood the issues and was excited by them. She'd get back to me. The next day I got a phone call. The exec sounded a little apologetic. "The story is great and we're really interested," she assured me, "but the guys here have a problem with the body building. They don't think it's sexy enough. How would you feel about making the woman a champion high diver? That way we could have her in a bathing suit and see lots of her body and have lots of other sexy women parading around swimming pools in small suits. If you agree to that change, we'll go into development."

She was dead serious. I sucked in my breath and tried not to laugh. "I'll call Zelda and ask her," I said and hung up the phone. I had a decision to make. I could call Zelda and convince her that she should change her sport from body building to high diving. I would have to work especially hard to do that but she might go for it if I told her she should. But I'd feel rotten doing that. Or, I could refuse to make the changes and lose the deal.

I thought hard about my decision. I knew Zelda wanted desperately to have a movie made about her life. She was broke and needed the money. She also needed the confidence boost the project and its publicity would give her. As far as I was concerned, body building was the major issue in the movie, literally and symbolically. Body building was like sculpture and the sport concentrated entirely on the physique, whereas diving concentrated not on the body, which was a tool, but on the performance of the body as it executed certain moves. There was a big difference here in focus and perspective, in subtext and context. Body building also involved the question of a definition of femininity as it related to the traditional masculinity of the sport. Diving had no such issues. By changing the sport, the production company really was changing the entire spirit of the movie.

I decided that the ethical thing to do was to tell Zelda what I thought, and what the production company wanted to do and to let her decide. I did that and, to her credit, in spite of her "desperation," she decided she wouldn't be happy with the kind of change the production company wanted. Diving wasn't about Zelda. Body building was.

Even though I retained the right to change the story in spite of what Zelda might say, I decided not to do that. I felt I couldn't make the points I wanted to make without completely ignoring everything Zelda had worked so hard to establish and why she wanted to sell her story in the first place. And I felt that making a sale wasn't as important as respecting someone's life and work.

I fought hard for what had now become "our" point of view, but the production company didn't see it our way. They passed on the project. Years later, other movies came out about bulimia. Mine would have been much much better, but never for a moment was I sorry that I didn't gut Zelda's story in order to get it made.

Imagine how difficult it would be if the person whose life story you were trying to sell was someone very close to you. Ask yourself if you'd be comfortable with negative ways in which you might have to portray them or ways in which you will have to move off the story that inspired you in the first place in order to get a movie made. And that question might throw you back into thinking about what motivates you to write screenplays.

Of course, if we're motivated by specific stories, we have to be true to those stories or we can't be true to ourselves. Too many writers, in order to sell their projects, are willing to move off the stories they should be fighting for. But it's a hard thing to do, particularly in pitch sessions when the pressure and tension of the moment often force you to dance with the punches and change your determined ways. Sometimes in a pitch, you'll want to insist that your film deliver a certain message and lots of times the people you are pitching to aren't comfortable with that message or don't agree with it.

During those times, it's a test of your mettle as to how to deal with the situation. Will you move off your message or change it to please the executives? Will you fight for your pitch at the risk of getting thrown out on your ear? It's a tough choice and one that sinks lots of screenwriters.

Now that doesn't mean that the pitch has to be obviously "messagey." It's quite alright to hide the "moral" when pitching the story. Sometimes it's even necessary. I found this out when I pitched a story I thought was a real winner because it was so inspirational. I began my pitch by telling the production executive that my film was about the triumph of the human spirit. That's all I got to say because the exec interrupted me by sitting forward in her chair and jabbing her finger in my face.

"Forget it!" she snapped. "I'm not interested in the human spirit! I'm only interested in sex!" Naturally my "relationship" with that particular executive was over, and it proves my point that most executives, like most audiences, don't want to be told the message. They may finally get it in a roundabout way, but going in they usually don't want to know what they should be thinking or how they should react. They want a good story and in the process, if that story is compelling, they'll go along with the message it delivers. "Hiding" the message in the pitch is not an unethical thing to do.

It's important to know when to compromise and when to stand firm. Certain things aren't worth fighting for and you've got to know what those are. In any story, think about the points you'd be willing to sacrifice to tell it. That's a hard thing to do because when you're involved in a story, every part of it seems important, but you might try the following.

Exercise

Rethink the story you want to write in the following ways:

- Change the sex of your main character (male lead to female lead) without changing the facts of your story.
- Change the location of your story.
- Change the tone of the story. See how dramas might become comedies and comedies might become dramas.

∾ Change the genre. Turn an action–adventure into a period piece. Turn a romantic comedy into a horror movie. This is can be fun and can lead to exciting new writing experiences.

∾ Change the profession of the hero and see how it influences the story. For example, if it's a lawyer story, make the main charac-ter a bricklayer.

∾ Change the time period. Turn a Western into a mid-west subur-ban story.

All these changes don't mean that you actually have to rewrite your script. It just means that you have to see how far you would be willing to go to make changes. This will come in particularly handy during pitch sessions in which executives can suggest some of the most in-sane changes imaginable.

For example, someone I know pitched a story about an American kid coming to live with his Irish grandfather in Ireland. The executive loved the story but wanted the writer to turn the script into a story about a black kid and his grandfather living in the South. Big change. Completely different story. Did the writer do it?

Yes. She wanted to make the sale! That decision was hers to make and wasn't wrong or right. It was simply what she thought she could live with. Of course, once she got involved in the writing process, she rued the day and cursed her choice, but by then it was too late. She had let her money motivation take her over. Because her heart wasn't in the script she finally wrote, it didn't succeed and was rewritten by another writer who wanted the opportunity to write about a black kid and his black grandfather living in the South.

What it finally all comes down to is courage—the courage to dis-cover who you are, the courage to make a choice for what your heart believes, the courage to stick with that choice.

COURAGE

ourage is not an absence of fear. People who are courageous often tell you that at the moment when courage compelled them to act, they were very afraid. Courage is the ability (and determination) to act in spite of fear, sometimes even because of it. People who act courageously do not allow themselves to be paralyzed by their fear, and their will power and determination make all the difference.

That's why courage requires strength—not physical strength, but strength of feeling and strength of character. That strength of character usually comes from a commitment to truth and to a higher purpose. Acts of courage need not be large. Sometimes even small and routine acts are courageous. Expressing an opinion, confronting a friend, going to work every day—all these things can take strength of character and courage.

Courage is as necessary in writing as it is in life. Writers who are courageous will dare to create things that are beyond their self-perceived limitations. They will go beyond boundaries of conventionality, and sometimes even of common sense. And they will do so relentlessly and ruthlessly in search of what for them is truth.

Sometimes it seems as if evil people are more courageous than good people. They seem to go further and to fight harder than good people do, but this is an illusion. I believe that evil people only seem to be more courageous at first, because they've saved their strength by being cowardly where it counts.

They haven't dared to take the time and trouble to explore the good within themselves and they haven't done the very hard work of making the commitment to bringing out that good. Evil people manifest a showy, superficial strength that doesn't last. Only people who have steadily built up a strength sustained by truth and goodness can go the distance to victory. That's why the star of the show always wins.

Evil doesn't demonstrate courage. It only demonstrates bravado and hubris. Real courage comes from standing against evil by holding to good even in the face of flagrant temptation to do otherwise.

226

It takes courage to stand against the norm, to champion or even express unpopular or unfashionable ideas. And sometimes, this courage may seem to go unrewarded. Courageous people sometimes fail in their efforts or even die because of them. And yet, because the rewards of courage are sometimes very subtle (peace of mind that comes from a clear conscience in knowing that you've done the right thing), courageous people seldom complain.

I believe that acts of courage are inevitable in the light of our own humanity if we are familiar and intimate with that humanity and love it as we love ourselves. Because I believe that courage takes practice and comes from the inspiration of self-discovery, here are some exercises you can do to build up your own courage.

Exercise

1. Gather a list of quotes about courage that mean something to you and put them in prominent places around your writing space.
2. Make a list of things you are "afraid" to do in order of magnitude and then, one by one, do them. Start with small things and then work up. Make sure that none of the things on your list is reckless or dangerous. If you are afraid of heights, for example, don't instantly start mountain climbing. Build up to facing your greatest fears. When we face our fears, they tend to disappear.

Courage, as it pertains to screenwriting, is very specialized indeed. Here are some ways in which screenwriters need to be courageous. You can add to the list if you want to.

Thirty Kinds of Screenwriting Courage

෴ Courage to face the blank computer screen and get to work.

෴ Courage to find out what you really think and then say it.

෴ Courage to commit to an idea and/or a story.

෴ Courage to expose your inner self to yourself.

෴ Courage to expose your inner self to others.

෴ Courage to write a crappy first draft because (as Hemingway said), "The first draft of anything is always shit."

෴ Courage to start another rewrite no matter how many drafts you've written.

∽ Courage to take chances with unfamiliar genres, techniques, and styles.

∽ Courage to stand behind your work.

∽ Courage to look stupid to yourself and to others.

∽ Courage to be unpopular.

∽ Courage to be uncommercial.

∽ Courage to be commercial.

∽ Courage to make it big and still remain true to your own voice.

∽ Courage to be rejected and still believe in your own voice.

∽ Courage to learn from your bad experiences and then "forget" them.

∽ Courage to admit mistakes and correct them.

∽ Courage to try and understand yourself and your own motives.

∽ Courage to love and forgive yourself in spite of your motives.

∽ Courage to understand other people and their motives.

∽ Courage to love and forgive other people in spite of their motives.

∽ Courage to love your work.

∽ Courage to hate your work.

∽ Courage to admit others know more than you do.

∽ Courage to take criticism and advice.

∽ Courage to act on or turn down that criticism and advice.

∽ Courage to act ethically no matter what the circumstances.

∽ Courage to listen to your conscience and live by what it says.

∽ Courage to endure.

∽ Courage to never give up.

   If you can master even one of these "courages" you'll find your life and your work changing more than you ever imagined it would.

CONCLUSION

AN ETHICS CHECK LIST

ow that you've absorbed all the information in this book and worked the exercises, you can use this ethics checklist in the development, pre-writing, and writing of each one of your screenplays.

- My screenplay idea is, to the best of my knowledge, original, and not "borrowed" from another source. If it's based on a true story or a book, I've acquired the rights to that story or to that book or made sure it's in the public domain.
- My screenplay idea is based on a theme I care about deeply.
- I have made a brief list of several things I want to say about this theme.
- I have educated myself about what I want to say by introspection, research, and discussion with others I respect.
- I have considered the social impact of my "message." I stand behind it and am willing to take the responsibility for delivering it in a popular medium.
- If sex and/or violence are present in my screenplay, they are there as an integral part of the story and add to character development and are not gratuitous.
- I have honored the process of decision making in my screenplay by demonstrating that each individual has the power to make a choice between good and evil, even though that choice may be difficult, or coerced by circumstances.
- My depiction of Good and Evil is based on what I truly believe and not on what sells.
- When I (or my characters) cannot make a clear distinction between what is good and evil, I make this "confusion" an integral

issue in the script rather than trying to get away with blurring my screenplay's definition of good and evil.

- The characters I've created have been motivated by deep inner issues rather than behaving like cardboard cutouts or stereotypes.

- My characters have performed deeds commensurate with their natures and not out of a need to shock or confound audiences.

- The "punishment" I've meted out to villains in my movie fits the crime—that is, I've not been outrageous (too lenient or too brutal) just for the sake of being outrageous.

- In order not to alienate audiences, I've presented my "message" in a non-preachy way.
  - My dialogue is short and pithy.
  - I seldom (if ever) let characters make on-the-nose pronouncements about the movie's message.
  - My plot progress depends on character motivation and not solely on the necessity to get the message across.

- In years to come, I will be able to look at what I've written and, without making excuses, be proud to claim it as part of my body of work.

ONLY THE BEGINNING

Writing this book has been a magnificent and exciting adventure. It took a long time to convince people that it should be written—that ethics should apply to creative work and that creative people needed to learn how to be ethical. Nearly every time I told someone I was writing a book about ethics for screenwriters, I got big laughs and/or strange looks. And over and over again I heard the line "Ethics for screenwriters? Isn't that an oxymoron?" I put all those reactions down to the disregard people have developed for showbiz and the people in showbiz. I think that's sad. But, with the world changing as it has since September 11th, 2001, with more and more people wanting to live lives of meaning and hoping to put "quiet desperation" behind them, with more and more film schools cropping up around the world filled with young and eager people not yet jaded by crass commercialism, I believe there is hope. I guess I always have. That's why I chose to write the kinds of things I've always written and why I chose to teach screenwriting and why I chose to write this book.

I believe that there are techniques we can use to become more spiritual, more thoughtful, more insightful, and more ethical. I even believe that if we try hard and stay committed, we can inspire our colleagues and the industry to make similar efforts. I hope this book has made a start. But in the great scheme of things, it's only the beginning. It's up to each one of us to do our part by changing ourselves. If we do that, then we truly will change the world.

Marilyn Beker
Los Angeles, California, 2002

233

ACKNOWLEDGMENTS

I am profoundly grateful to those whose presence in my life gave me the confidence to take on the subject of this book: the saints and sages of all religions—compassionate world teachers who loved humanity enough to share their hard-won knowledge—most particularly Paramahansa Yogananda; J. Oliver Black (Yogacharya Oliver), who taught me the importance of the practical application of spiritual teachings; my sister, Jeannie Beker, for her inspiring example of unflagging courage, generosity, and honesty. Besides all that, the stories she tells me about her jet-set lifestyle always put the world into perspective and make me laugh; my niece Becky for her love of literature and her artistic flare, and my niece Joey for her love of art and her sense of comedy; Bill Barclay, dear to my heart and there at the beginning, now and always; my divine friends during the early days of Song of the Morning Ranch (among them Deborah Carlson, Irmgard Kurtz, Carol Harder, Jan Sheetz, Jane O'Brien, Sue Brown, Jeff Duncan, Chris Kurtz Duncan, Brij Chhabra, Roy Thibodeau, Mitch Kamiel, Deanna Kamiel, Ian Wylie, Richard Armour, Carol Sirosky Armour, Rosemary Massey)—the memory of those friendships and the magic of the Ranch continue to sustain me to this day; my Los Angeles family, Ruth, Allan, Chana, and Noah Ickowitz—exceptional people whose constancy, love, and caring mean the world to me; Margie Friedman for her longtime loyal friendship; buddies Gillian Barnes and Steve Morris for their encouragement; Greg Frank, Hymie, Amy and Maggie Milstein, Darris, Irene, and now, Juliette Gringeri, for their love and support; Richard Freiman for his legal expertise and his camaraderie; fur-faces Betty, Riley, Muse, and Spense for their example of innocence and unconditional love; Dr. Hedda Bolgar, Dr. Richard Metzner, Dr. Lew Richfield, Dr. Joel Bienenfeld, and Dr. Doris Owanda Johnson for helping me to sustain clarity and energy; Loyola Marymount University, Dean Emeritus Tom P. Kelly, Dr. Joe Jabbra, and Howard Lavick for their early support of this work; Kelly O'Hara for her careful reading of parts of this manuscript; my brilliant editor, Linda Bathgate, who had the vision it took to give a book like this book a home and made its writing a wonderful experience; and to those former students—now friends—of whom I am very proud (among them Vanessa M. Coto, Tanya Davies, Jason Endres, Darris Gringeri, Bob Herbstman, Kelly O'Hara, Fr. Luis Proenca S. J., Whitney Zaring) who pay me the continuing compliment of taking what I say to heart and living and writing ethically.

REFERENCES

Arnold, Sir Edwin. (Trans.). (1934). *The song celestial or Bhagavad-Gita.* Philadelphia: David McKay.

Beker, Marilyn. (1989, March/April). David Cronenberg. *Expression Magazine.*

Bradbury, Ray. (1989). *Zen in the art of writing.* Santa Barbara, CA: Capra Press.

Christians, Clifford G., Facler, Mark, & Rotzoll, Kim B. (1997). *Media ethics: Cases and moral reasoning* (5th ed.). New York: Longman.

Dawood, N. J. (Trans.). (1997). *The Koran.* London: Penguin Books.

Diamond, Nina L. (2001, July 29). True believer. *The Los Angeles Times Magazine.*

Dalai Lama. (1999). *Ethics for the new millennium.* New York: Riverhead Books.

Emerson, Ralph Waldo. (1951). *Essays by Ralph Waldo Emerson.* New York: Crowell.

Eller, Claudia. (2001, September 14). Hollywood executives rethink what is off limits. *The Los Angeles Times.*

Froug, William. (1991). *The new screenwriter looks at the new screenwriting.* Los Angeles: Silman-James Press.

Goldstein, Robert. (2001, September 25). A turn of events, a change in plot. *The Los Angeles Times.*

Gordon, Devin. (2001, November 19). Mamet's on a classic caper. *Newsweek Magazine.*

Hanayama, Shoyu (Ed.), & Steiner, Richard R. (Trans.). (2001). *The teaching of Buddha.*
Tokyo: Bukkyo Dendo Kyokai.

Harris, Erich Leon. (1966). *African American screenwriters now.* Los Angeles: Silman-James Press.

Hohenadel, Kristin. (2001, July 1). Film goes all the way in the name of art. *The New York Times.*

Holy Bible. (1952). New York: Thomas Nelson & Sons.

Keller, Helen. (1974). *My religion.* New York: Pyramid Books.

Maritain, Jacques. (1960). *The responsibility of the artist.* New York: Scribner's.

May, Rollo. (1978). *The courage to create.* New York: Bantam Books.

Miller, James E. (Ed.). (1959). *Walt Whitman: Complete poetry and selected prose.* Boston: Houghton Mifflin.

Paramahansa Yogananda. (1969). *The science of religion.* Los Angeles: Self-Realization Fellowship Press.

Paramahansa Yogananda. (1999). *A world in transition: Finding spiritual security in times of change.* Los Angeles: Self-Realization Fellowship Press.

Plimpton, George. (Ed.). (1967). *The Paris Review interviews writers at work* (3rd series). New York: Penguin Books.

Plimpton, George. (Ed.). (1984). *The Paris Review interviews writers at work* (6th series). New York: Penguin Books.

Pope John Paul II. (1997). *Crossing the threshold of hope.* (Vittorio Messori, Ed.). New York: Knopf.

Rodley, Chris. (Ed.). (1993). *Cronenberg on Cronenberg.* London: Faber & Faber.

Ruas, Charles. (1985). *Conversations with American writers.* New York. Knopf.

Salwen, Michael, & Stacks, Don. (Eds.). (1996). *An integrated approach to communication theory and research.* Mahwah, NJ: Lawrence Erlbaum Associates.

Sissen, C. H. (Trans.). (1993). *The divine comedy.* New York: Oxford University Press.

Smith, Lynn. (2001, November 17). Does Hal send mixed signals? *The Los Angeles Times.*

Snyder, H., & Sickmund, M. (1999). *Juvenile offenders and victims: 1999 national report.* Washington, DC: Office of Juvenile Justice and Delinquency Prevention.

Solzhenitsyn, A. (1980). *The oak and the calf.* New York: Harper & Row.

Sri Daya Mata. (1990). *Finding the joy within you.* Los Angeles: Self-Realization Fellowship Press.

Stayton, Richard. (1999, June/July). Steve Martin. *Written By Magazine,* 6.

Taylor, Harold. (1960). *Art and the intellect.* New York: The Museum of Modern Art/Doubleday.

Tolstoy, Lev Nikolaevich. (1932). *What is art?* London: Oxford University Press.

Tolstoy, Lev Nikolaevich. (1948). *The law of love and the law of violence.* New York: Rudolph Field.

Wichler, Stephen E. (Ed.). (1960). *Selections from Ralph Waldo Emerson, an organic anthology.* Boston: Houghton Mifflin.

Wolfe, Alan. (2001, March 18). The final freedom. *The New York Times Magazine.*

INDEX